"This book is a delight and its approach long overdue. By broadening its interpretation of the meaning of 'development' and by addressing many of the most contested issues in academic and political debates in and on Latin America, it brings home to the reader, but in an eminently comprehensible and digestible form, the complexity of the region's recent and contemporary patterns of economic, social and political transformation. Moreover, to its great credit, it does so with a sharp eye, a sympathetic ear and a clarity and comprehensiveness which have all been sadly lacking in such books for several decades. This book sets the picture straight and does us all an invaluable service."

Antoni Kapcia, Professor in Latin American History,
Faculty of Arts, University of Nottingham, UK

"This book is a critical introduction to Latin America and its experience with development, and provides an invaluable resource to both students and scholars. Historically and geographically rich, this book examines processes of state formation and political economic thought through the colonial period and the Republican era, the emergence of Industrial Import Substitution and the dependency school, and the rise of neoliberalism and its discontents. These historical transformations have of course been marked by ideological struggle and political violence, and Julie Cupples proves the ideal guide to the continent's shifting political and economic landscapes. The book highlights difference as well as continuity, and its attention to the region's immense ethnic, linguistic, cultural, class, and ecological diversity is particularly valuable. *Latin American Development* provides a nuanced and authoritative look at political and economic change in the American continent."

Tom Perreault, Associate Professor of Geography,
Department of Geography, Syracuse University, USA

Latin American Development

Latin America's diverse political and economic struggles and triumphs have captured the global imagination. The region has been a site of brutal dictators, revolutionary heroes, the Cold War struggle and a place in which the global debt crisis has had some of its most lasting and devastating impacts. Latin America continues to undergo rapid transformation, demonstrating both inspirational change and frustrating continuities.

This text provides a comprehensive introduction to Latin American development in the twenty-first century, emphasizing political, economic, social, cultural and environmental dimensions of development. It considers key challenges facing the region and the diverse ways in which its people are responding, as well as providing analysis of the ways in which such challenges and responses can be theorized. The book also explores the region's historical trajectory, the implementation and rejection of the neoliberal model and the role played by diverse social movements. Relations of gender, class and race are considered, as well as the ways in which media and popular culture are forging new global imaginaries of the continent. The text also considers the increasing difficulties that Latin America faces in confronting climate change and environmental degradation.

This accessible text gives an overarching historical and geographical analysis of the region and critical analysis of recent developments. It is accompanied by a diverse range of critical historical and contemporary case studies from all parts of the continent, providing readers with the conceptual tools required to analyse theories on Latin American development. Each chapter ends with a summary section, discussion topics, suggestions for further reading, websites and media resources. This is an indispensable resource for scholars, students and practitioners.

Julie Cupples currently lectures in Human Geography at the University of Edinburgh. She has been working in and on Latin American development for more than two decades, through both Central American solidarity and academic scholarship. Most of her research has been carried out in Nicaragua. She has published widely on a range of themes including postdevelopment, gender, revolution and conflict, neoliberalism, elections, municipal governance, indigenous and community media, and climate change, disasters and environmental risk.

ROUTLEDGE PERSPECTIVES ON DEVELOPMENT

Series Editor: Professor Tony Binns, *University of Otago*

Since it was established in 2000, the same year as the Millennium Development Goals were set by the United Nations, the *Routledge Perspectives on Development* series has become the pre-eminent international textbook series on key development issues. Written by leading authors in their fields, the books have been popular with academics and students working in disciplines such as anthropology, economics, geography, international relations, politics and sociology. The series has also proved to be of particular interest to those working in interdisciplinary fields, such as area studies (African, Asian and Latin American studies), development studies, environmental studies, peace and conflict studies, rural and urban studies, travel and tourism.

If you would like to submit a book proposal for the series, please contact the Series Editor, Tony Binns, on: jab@geography.otago.ac.nz

Published:

Third World Cities, 2nd edition
David W. Drakakis-Smith

Rural–Urban Interactions in the Developing World
Kenneth Lynch

Environmental Management and Development
Chris Barrow

Tourism and Development
Richard Sharpley and David J. Telfer

Southeast Asian Development
Andrew McGregor

Population and Development
W.T.S. Gould

Postcolonialism and Development
Cheryl McEwan

Conflict and Development
Andrew Williams and Roger MacGinty

Disaster and Development
Andrew Collins

Non-Governmental Organisations and Development
David Lewis and Nazneen Kanji

Cities and Development
Jo Beall

Gender and Development, 2nd Edition
Janet Momsen

Economics and Development Studies
Michael Tribe, Frederick Nixson and Andrew Sumner

Water Resources and Development
Clive Agnew and Philip Woodhouse

Theories and Practices of Development, 2nd Edition
Katie Willis

Food and Development
E.M. Young

An Introduction to Sustainable Development, 4th Edition
Jennifer Elliott

Latin American Development
Julie Cupples

Religions and Development
Emma Tomalin

Forthcoming:

Global Finance and Development
David Hudson

Natural Resource Extraction and Development
Roy Maconachie and Gavin M. Hilson

Development Organizations
Rebecca Shaaf

Children, Youth and Development, 2nd Edition
Nicola Ansell

Climate Change and Development
Thomas Tanner and Leo Horn-Phathanothai

Politics and Development
Heather Marquette and Tom Hewitt

South Asian Development
Trevor Birkenholtz

Conservation and Development
Shonil Bhagwat and Andrew Newsham

Latin American Development

Julie Cupples

LONDON AND NEW YORK

First published 2013
by Routledge
2 Park Square, Milton Park, Abingdon, Oxon OX14 4RN

Simultaneously published in the USA and Canada
by Routledge
711 Third Avenue, New York, NY 10017

Routledge is an imprint of the Taylor & Francis Group, an informa business

British Library Cataloguing in Publication Data
A catalogue record for this book is available from the British Library

Library of Congress Cataloging-in-Publication Data
Cupples, Julie.
Latin American development / Julie Cupples.
pages; cm
Includes bibliographical references and index.
1. Latin America–Social conditions–21st century.
2. Latin America–Economic conditions–21st century. I. Title.
HN110.5.A8C87 2013
306.098–dc23
2012039212

ISBN: 978–0–415–68061–5 (hbk)
ISBN: 978–0–415–68062–2 (pbk)
ISBN: 978–0–203–55486–9 (ebk)

Typeset in Times New Roman
by Keystroke, Station Road, Codsall, Wolverhampton

Printed and bound in Great Britain by
TJ International Ltd, Padstow, Cornwall

For Sadie Rivas (1962–1999)

 # Contents

Illustrations

Figures

Table

 **Boxes**

Acknowledgements

In many ways, this book has been a long time in the making. My interest in Latin American development began when I was an undergraduate student of Spanish at the University of Bradford in the mid-1980s. Ronald Reagan was president of the United States and I became fascinated by the Nicaraguan Revolution and the struggle to defend it against US imperialism. I became involved in Nicaraguan and Central American solidarity, a movement through which I got to know and work with many inspirational people in both Central America and the United Kingdom. I made my first visit to Nicaragua in 1990 and became totally hooked on the courageous and creative political struggles I witnessed; these struggles have continued to be the basis of my activism and my academic career.

Consequently, many people have been fundamental in this journey and have shaped my thinking in so many ways. In Nicaragua, I would like to thank above all Sadie Rivas of Matagalpa. Sadie was a guerrilla fighter, political activist and development practitioner, who was tragically taken from us in a car accident during my doctoral fieldwork in 1999. Sadie embodied the best of Latin American development. Her life and struggle inspired my PhD thesis and continue to motivate me to be part of the attempt to create a better world in Central America.

In addition, I would like to thank the Instituto de Investigaciones de Gestión Social (INGES) in Managua and especially my dear and inspirational friend and collaborator, Irving Larios. Thanks also go to members of the Communal Movement of Matagalpa, especially Auxiliadora Romero, Sergio Saénz, Janett Castillo and Zoila Hernández and a number of Creole and Miskito civil society leaders, intellectuals and mediamakers on Nicaragua's Atlantic Coast, especially Dixie Lee.

Acknowledgements also go to the Carter Center in Atlanta, Georgia, for having invited me to be part of two electoral missions to Nicaragua. For their political support and friendship through the UK-based Nicaragua Solidarity Campaign, I would like to thank in particular Anne Carruthers, Guy Pilkington, Emma Dooks, Bob Thorpe, Janet Toland, Martin Toland, Clive Davies and José Lara.

In New Zealand, I would like to thank the Department of Geography of the University of Canterbury, where I worked for a decade from 2002 to 2012, for substantial financial support for fieldwork in Central America which has culminated in a number of publications. I would like to thank Marney Brosnan at the University of Canterbury for her genuine interest in Latin American development and the substantial audiovisual, cartographic and fieldwork support she has provided over the past 10 years, including the supply of photos and the map for this book. For supervision, mentoring, professional invitations, collaborations and many fruitful conversations about Latin American development, I would also like to thank Sara Kindon, Audrey Kobayashi, Wendy Larner, Nina Laurie, Eric Pawson, Marcela Palomino, Richard Peet, Regina Scheyvens and James Sidaway. Finally, I would like to thank my beautiful children, Natasha and Ruben Vine, for accompanying me on parts of this journey and for the way in which as young children they embraced life in Nicaragua; and my husband and co-author, Kevin Glynn, whose immense emotional and intellectual input into my work can never be adequately acknowledged.

 **Abbreviations**

ACPO	Acción Cultural Popular – Cultural Popular Action (Colombia)
ALBA	Alianza Bolivariana para los Pueblos de Nuestra América – Bolivarian Alliance for the Peoples of Our America
ALN	Ação Libertadora Nacional – National Liberation Action (Brazil)
AMD	Acid mine drainage
AMNLAE	Asociación de Mujeres Nicaragüenses Luisa Amanda Espinoza – Association of Nicaraguan Women Luisa Amanda Espinoza
APPO	Asamblea Popular de los Pueblos de Oaxaca – Popular Assembly of the Peoples of Oaxaca (Mexico)
APRA	Alianza Popular Revolucionaria Americana – American Popular Revolutionary Alliance (Peru)
BRIC	Brazil, Russia, India, China
CAFTA	Central American Free Trade Agreement
CCER	Coordinadora Civil para la Emergencia y la Reconstrucción – Civil Coalition for Emergency and Reconstruction
CDM	Clean Development Mechanism
CEB	Comunidades Eclesiástica de Base – Comunidades Eclesiais de Base – Christian Base Communities
CERJ	Consejo Étnico Runejel Junam – Runejel Junam Council of Ethnic Communities (Guatemala)
CIA	Central Intelligence Agency (US)
CIDOB	Confederación de Pueblos Indígenas de Boliviano – Confederation of Indigenous Peoples of Bolivia

CONAIE	Confederación de Nacionalidades Indígenas de Ecuador – Confederation of Indigenous Nationalities of Ecuador
CONAVIGUA	Comité Nacional de Viudas de Guatemala – National Committee of Widows of Guatemala
CPR	Comunidades de Pueblos en Resistencia – Communities of Peoples in Resistance (Guatemala)
CUC	Comité de Unidad Campesina – Committee for Peasant Unity (Guatemala)
CUD	Coordinadora Unica de Damnificados – Overall Coordinating Committee of Disaster Victims (Mexico)
DEA	Drug Enforcement Administration (US)
DREAM	Development Relief and Education for Alien Minors (US)
ELN	Ejército de Liberación Nacional – National Liberation Army (Colombia)
ENSO	El Niño Southern Oscillation
ERP	Ejército Revolucionario del Pueblo – Revolutionary Army of the People (Argentina)
EU	European Union
EZLN	Ejército Zapatista de Liberación Nacional – Zapatista Army of National Liberation (Mexico)
FARC	Fuerzas Armadas Revolucionarias de Colombia – Revolutionary Armed Forces of Colombia (Colombia)
FDI	Foreign Direct Investment
FMLN	Frente Farabundo Martí para la Liberación Nacional – Farabundi Martí Front for National Liberation (El Salvador)
FMSO	Foreign Military Studies Office (US)
FSLN	Frente Sandinista para la Liberación Nacional – Sandinista Front for National Liberation (Nicaragua)
FTA	Free Trade Agreement
FTAA	Free Trade Area of the Americas
FTZ	Free Trade Zone
GAD	Gender and Development
GAM	Grupo de Apoyo Mutuo – Group of Mutual Support (Guatemala)
GDP	Gross Domestic Product
GIS	Geographic Information Systems
GNI	Gross National Income
GPS	Global Positioning Systems
HDI	Human Development Index
HIPC	Heavily Indebted Poor Countries Initiative
HTA	Home Town Association

ICT4D	Information and Communication Technologies for Development
IMF	International Monetary Fund
ISI	Import Substitution Industrialization
MAS	Movimiento al Socialismo – Movement for Socialism (Bolivia)
MCD	Modernity/coloniality/decoloniality
MDRI	Multilateral Debt Reduction Initiative
MNR	Movimiento Nacionalista Revolucionario – Revolutionary Nationalist Movement (Bolivia)
MNU	Movimento Negro Unificado – Unified Black Movement (Brazil)
MST	Movimento dos Trabalhadores Rurais Sem Terra – Landless Workers Movement (Brazil)
MTD	Movimiento de Trabajadores Desempleados – Movement of Unemployed Workers (Argentina)
NAFTA	North American Free Trade Agreement
NIDL	New International Division of Labour
NGO	Non-governmental organization
NSM	New social movement
NTAE	Non-traditional agricultural export
OAS	Organization of American States (in Spanish OEA – Organización de Estados Americanos)
OCMAL	Observatorio de Conflictos Mineros de América Latina – Observatory of Mining Conflicts in Latin America
OPEC	Organization of Petroleum Exporting Countries
OPIP	Organización de Pueblos Indígenas de Pastaza – Organization of Indigenous Peoples of Pastaza (Ecuador)
PAN	Partido Acción Nacional – National Action Party (Mexico)
PCN	Proceso de Comunidades Negras – Process of Black Communities (Colombia)
PES	Payment for Ecosystem Services
PRD	Partido de la Revolución Democrática – Party of the Democratic Revolution (Mexico)
PRI	Partido Revolucionario Institucional – Institutional Revolutionary Party (Mexico)
PRSP	Poverty Reduction Strategy Paper
PT	Partido dos Trabalhadores – Workers Party (Brazil)
RCN	Radio Cadena Nacional (Colombia, television channel)
REDD	Reducing Emissions from Deforestation and Degradation

SERNAM	Servicio Nacional de la Mujer – National Service for Women (Chile)
SICA	Sistema de Integración Centroamericana – Central American Integration System
TIPNIS	Territorio Indígena y Parque Nacional Isiboro Secure – Isiboro Sécure National Park and Indigenous Territory (Bolivia)
UFCO	United Fruit Company
UN	United Nations
UNASUR	Unión de Naciones Suramericanas – União de Nações Sul-Americanas – Union of South American Nations
UNDP	United Nations Development Programme
URNG	Unidad Revolucionaria Nacional Guatemalteca – Guatemalan National Revolutionary Unity (Guatemala)
VPR	Vanguarda Popular Revolucionária – Revolutionary Popular Vanguard (Brazil)
WAD	Women and Development
WDR	World Development Report
WGIP	Working Group on Indigenous Populations
WID	Women in Development
WTO	World Trade Organization
WWF	World Wildlife Federation

1 Introduction

What/where is Latin America?

Learning outcomes

By the end of this chapter, the reader should:

- Understand the complexities that surround defining Latin America as a region
- Develop a sense of Latin America's geographic diversity
- Gain a preliminary insight into the some of the main theoretical approaches used to study Latin American development

(Latin) America

Studying Latin American development means studying the politics of contestation. Since the time of independence from European powers, Latin America has experimented with a range of development models, and the continent has both produced and been subject to a variety of economic, political and cultural thought concerned with the development question. Indeed, "development" as a concept and the modes of sensemaking that have coalesced around it are central to the emergence and consolidation of Latin America as a region. Latin America's historical record provides some of the world's most fascinating struggles between states and markets, between liberals and conservatives, between capitalism and socialism, between sustainability and environmental destruction, between democracy and dictatorship, and between different understandings about race and gender. Latin America has proved to be a challenging, rewarding and multifaceted development education for many scholars, practitioners and activists.

Studying Latin American development also means studying the politics of globalization. The so-called discovery of America by Christopher Columbus and other European conquistadores in the fifteenth and sixteenth centuries constituted an early exercise in globalization, and since that period Latin America has both contributed to and been impacted by diverse globalizing processes, in ways that are positive, negative and contradictory. Latin America is a region that has long captured the global imagination. Since the 1960s, Che Guevara has been a global symbol of political resistance (Figure 1.1). In the 1990s and 2000s, a similar status has come to be granted to Subcomandante Marcos of the Zapatistas, whose ideas have inspired people all over the world. People in Russia and Bosnia are avid fans of Venezuelan and Brazilian telenovelas (Martínez 2005) and salsa is the most popular dance class in London (Rendell 2011). Many Latin American development issues, including the debt crisis of the 1980s, the deforestation of the Amazon rainforest and the human rights abuses committed by the military governments and dictatorships of the 1970s and 1980s, have attracted global political concern and activism.

Many of these issues are entangled with Latin America's historically complicated relationship with the United States. The US has tended to view Latin America as its "backyard" – an attitude that has resulted in military interventions, support for *coups d'état* and the imposition of economic policies and trade agreements that favoured US interests (see Livingstone 2009). The Cold War, ostensibly about the geopolitical struggles between the United States and the Soviet Union, was played out in Central America, with tragic consequences. At the present time, Mexican and Central American people perish with alarming frequency in the deserts on the US–Mexican border, in their attempt to enact a form of globalization wrapped in the promise of development.

Migratory, cultural and commercial flows, and the movement of Latin American people, icons, commodities, cultural products, ideas and discourses around the globe, mean that Latin America is constantly exceeding its borders. The dramatic growth of the Latin American population overseas, especially in the United States, is significant. There are now more than 50 million people of Latin American descent in the United States, and 7 million of them reside in Los Angeles (Ennis, Ríos Vargas and Albert 2011). There are also 1.8 million Latin Americans in Spain (EFE 2008), 240,000 Latin Americans in Canada (Lindsay 2007) and 130,000 Latin Americans in the UK (Linneker and McIlwaine 2011). These demographic changes are stretching the boundaries of Latin America, in former colonial centres and in parts of the contemporary US that were once part of Mexico, and the remittances sent back to family members at home far exceed foreign aid.

Studying Latin American development also means studying the cultural politics of hybridity and fusion. Latin America is often described as a meeting of three worlds – the European, the African and the native

Figure 1.1 *Mural dedicated to Latin American revolutionary Che Guevara, in Belfast, Northern Ireland*

Source: Ardfern, Wikimedia Commons (Creative Commons Attribution-Share Alike 3.0 Unported).

American – a mixing that continues to shape everyday life in the region. Although the indigenous and African populations were subjected to unimaginable cruelty, including genocide, many of their cultural forms and practices have survived.

Since the colonial era, Latin Americans have tactically and selectively blended indigenous and African cultures with European ones. Despite the undeniable loss of cultural memory and tradition inflicted on indigenous and African heritages, these cultures continue to assert themselves in the development project, through popular culture, religious practices, political struggles and everyday life. They continue to hybridize European-derived and imposed cultural forms and thus complicate what we mean by development. Even Catholicism, imposed on the native people by the conquistadores, was hybridized and therefore controlled (Rowe and Shelling 1991) through blending with pre-Columbian and African religions. Contemporary devotion to Candomblé, santería and the Santa Muerte is testament to the cultural resilience of non-European religious practices and their survival over time and against the odds.

In recent decades, through the liberation theology movement, Latin America has also used Catholicism to reflect on and challenge political injustice. When reflecting on hybridization, it is important not to lose sight of the fact that Latin Americans, like Europeans and Australians, are very receptive to US cultural influences and consume US cultural products such as music, Hollywood movies, fast food, and Coca-Cola with regularity, and many Latin Americans are joining US-style Evangelical churches. So US cultural influences, as well as economic and political ones, for better or worse, are also part of the mix (see Figure 1.2).

As Argentinian theorist Néstor García Canclini (2002) writes from his adoptive home in Mexico City, talking of Latin America as a whole is not easy. As he points out, in any attempt to compare Argentina and Mexico, the divergences often overshadow the similarities. It seems like a cliché to say that Latin America is a continent of contrasts, but there is no doubt that conceptualizing such a large and diverse region as a single entity is a difficult task. The continent as a whole confounds any straightforward generalizations about the region's development, given the stark and intimate ways that affluence and poverty, and modernity and tradition, are juxtaposed. A few minutes spent poring over the development indicators in Table 1.1 reveals as many differences and divergences between Latin American countries as similarities. There are also stark economic and

Figure 1.2 *Downtown Buenos Aires, Argentina: multinational corporations assert their presence on the urban landscape of Latin America*

Source: Natasha Vine.

cultural differences within countries, regions and neighbourhoods. All
Latin American cities, as well as some of its rural areas, are characterized
by dramatic socio-cultural heterogeneity. As García Canclini (2002)
writes, in Latin America we can observe the undeniable modernization of
cities such as Buenos Aires, Mexico City, Caracas, Santiago de Chile,
Bogotá and São Paulo with their corporate skyscrapers, immaculately
dressed business people and luxurious shopping malls (see Figures 1.3
and 1.4), which exists alongside poverty-induced social unrest, drug
trafficking, urban insecurity and hunger. The world's richest man, Carlos
Slim, is Mexican and there are affluent families in Montevideo who fly to
Miami in order to go shopping. There are also many Latin American
children who go to bed hungry every night and who do not attend school
because they do not own a pair of shoes. In terms of health, Latin
America has to deal with poverty-related illnesses and diseases such as
malnutrition, malaria, dengue and gastrointestinal diseases as well as
"modern" problems such as violence, HIV-AIDS and heart disease
(Franko 2007).

Figure 1.3 *The skyscrapers of downtown Caracas, Venezuela*

Source: Paulino Moran, Wikimedia Commons (Creative Commons Attribution 2.0 Generic).

Table 1.1 Selected development indicators

	Population: millions (WDR 2012)	HDI ranking (UNDP 2011)	Latin America HDI ranking (UNDP 2011)	Life expectancy, years (UNDP 2011)	Under 5 mortality rate per 1000 (World Bank 2009)	Mean years of schooling (UNDP 2011)	GNI per capita, US$ (World Bank 2010)	External debt, $US billions (World Bank 2009)	Internet users per 100 people (UN World Statistics Pocketbook 2009)	Percentage of population below the poverty line (CIA World Factbook)	GINI index (World Bank 2008)	Gender inequality ranking (UNDP 2011)	CO_2 emissions per capita, tonnes (UN World Statistics Pocketbook 2007)
Argentina	40	45(VH)	2	75.9	14	9.3	$8,450	$119	34.0	30 (2011)	46.3	67	4.7
Belize	0.3	93(H)	13	76.1	18	9.2	$3,740	$0.8	11.7	43 (2010)	n/a	n/a	1.4
Bolivia	10	108(M)	18	66.6	51	9.2	$1,790	$2.4	11.2	30 (2011)	56.3	88	1.4
Brazil	190	84(H)	11	73.5	21	7.2	$9,390	$260	39.2	26 (2008)	55.1	80	1.9
Chile	17	44(VH)	1	79.1	9	9.7	$9,940	$404	41.3	12 (2009)	52.1 (2009)	68	4.3
Colombia	46	87(H)	12	73.7	19	7.4	$5,510	$44	49.4	46 (2009)	57.2	91	1.3
Costa Rica	5	69(H)	7	79.3	11	8.3	$6,580	$7.5	32.4	16 (2006)	48.9	64	1.8
Cuba	11	51(H)	4	79.1	6	10.2	$5,437 (UN 2009)	–	14.3	n/a	n/a	58	2.4
Dominican Republic	9.8	98(M)	14	73.4	32	6.9	$4,860	$9.2	26.8	42 (2004)	49.0	90	2.2
Ecuador	14	83(H)	10	75.6	24	7.6	$4,510	$12	24.6	33 (2010)	50.6	85	2.2
El Salvador	6	105(M)	16	72.2	17	7.7	$3,360	$10	12.1	38 (2005)	46.8	93	1.0
French Guiana*	0.2	n/a	n/a	77	n/a	n/a	n/a	n/a	25.7	n/a	n/a	n/a	4.4
Guatemala	14	131(M)	22	71.2	40	4.1	$2,740	$12	16.3	56 (2004)	n/a	109	1.0
Guyana	0.76	117(M)	19	69.9	35	8.5	$3,270	$0.67	24.9	n/a	n/a	106	2.0
Haiti	10	158(L)	23	62.1	87	4.9	$650	$0.9	10.0	65 (2010)	n/a	123	0.3
Honduras	7.6	121(M)	20	73.1	30	6.5	$1,880	$1.8	9.8	18 (2011)	61.3	105	1.2

Mexico	112	57(H)	5	77	17	8.7	$9,330	$177	28.3	48 (2005)	48.3	79	4.4
Nicaragua	6	129(M)	21	74	26	5.7	$1,080	$2	3.5	26 (2010)	n/a	101	0.8
Panama	3.4	58(H)	6	76.1	23	9.4	$6,990	$11	27.8	19 (2009)	52.0 (2009)	95	2.2
Paraguay	6.4	107(M)	17	72.5	23	7.8	$2,940	$3.7	17.4	19 (2009)	52.1	87	0.7
Peru	29	80(H)	9	74	21	9.6	$4,710	$26	31.4	35 (2009)	49.0	72	1.6
Puerto Rico*	3.9	n/a	n/a	79.3	–	–	$16,105 (UN 2009)	n/a	25.1	n/a	n/a	n/a	n/a
Suriname	0.49	104(M)	15	70.6	26	7.2	$5,920 (2008)	$0.5 (CIA 2005)	31.4	70 (2002)	n/a	n/a	4.8
Uruguay	3.3	48(H)	3	77	13	8.4	$10,590	$11	41.8	21 (2001)	46.3	62	1.9
Venezuela	28	73(H)	8	74.4	18	6.2	$11,590	$55	31.2	38 (2005)	n/a	78	6.0

*Not an independent nation-state.

Notes :

The Human Development Index (HDI) measures development by combining life expectancy, educational achievement and income into a composite index. The scores enable the UNDP to rank countries as 1–47 (very high human development, VH), 48–94 (high human development, H), 95–141 (medium human development, M) and 142–187 (low human development, L). A country with perfect income equality would score 100.

The GINI index measures inequality according to the distribution of income on a scale from 1 to 100. A country with perfect income equality would score 1 whereas a country with maximal inequality (where one person has all the income) would score 100.

The Gender Inequality Index measures and ranks countries according to inequality in achievements between women and men according to three dimensions: reproductive health, empowerment and the labour market. The country with the highest gender equality is ranked 1.

Figure 1.4 *The Metrocentro, an upmarket shopping mall in Managua, Nicaragua*

Source: Vrysxy, Wikimedia Commons (Creative Commons Attribution 3.0 Unported).

It is just as important to consider the role played by the affluent families of Montevideo in Latin American development as that of rural malnourished children in the Guatemalan highlands. Inequality – the difference between rich and poor – is as much a development issue as poverty is. Attending to Latin American development means thinking through wealth and affluence, a point that many development practitioners with their understandable focus on poverty tend to overlook. In all large Latin American cities, there are attempts to manage inequality through forms of urban segregation, which include high-security walls, gated communities and armed guards at banks and restaurants frequented by the wealthy. Despite these attempts, the rich and poor intermingle constantly. At the traffic lights in Managua, the capital of Nicaragua, poor malnourished children clean the windscreens of Toyota landcruisers belonging to Managua's wealthy in exchange for a few coins. The rich and poor are also brought into intimate proximity through the provision of labour services, as the poor inhabit the homes of rich on a daily basis to clean their homes, do their laundry, care for their children and tend their gardens.

Despite Latin America's dramatic socio-cultural heterogeneity, the region does have a shared history of colonization, pillage, dependency and exploitation, which makes a text on Latin America treated as a whole appropriate (see Radcliffe and Westwood 1996). In the early 1970s, Eduardo Galeano (1971), who is arguably Uruguay's most important political writer, described Latin America as a continent with open veins. It was brought into being as a geopolitical entity in order to service the needs of foreigners. Latin America's wealth became the source of its poverty, as foreigners earned more from consuming Latin America's oil, iron, copper, meat, fruit and coffee than Latin America did from producing them (Galeano 1971: 1). The role of supplier of raw materials in the service of development elsewhere is one that was created five centuries ago but continues to have repercussions today for the economic, political and cultural life of the region.

It is also important to interrogate the "Latin" in Latin America. While America (or América) includes Chile, Uruguay, Bolivia, Guatemala and Canada, for many people in the world America is now synonymous with the United States. As Galeano (1972) states:

> Along the way we have lost the right to call ourselves *Americans*, although the Haitians and the Cubans appeared in history as new people a century before the *Mayflower* pilgrims settled on the Plymouth coast. Now America is for the world nothing more than the United States; as a result where we live is a sub-America, a second-class America of nebulous identity.
>
> (emphasis in original)

It was not until the nineteenth century that the concept of Latin America was invented as a means of asserting difference from Anglo America, by both Latin Americans and Anglo Americans. McGuinness (2003) believes that the concept of "Latin America" gained momentum in the mid-nineteenth century as a result of the role and presence of white US American travellers and traders and the development of trading and transport routes which created the opportunity for everyday racialized encounters and exchanges. It was the time of the California gold rush, and the completion of a railway line across Panama made it the preferred route from the Atlantic to the Pacific. At this particular time, Panama had abolished slavery and was developing complex modes of racial democracy and national sovereignty which included the accordance of social and political status to people that white US Americans deemed to be racially inferior. In the encounters between Panamanians and white US Americans, the latter frequently attempted to impose racist and exclusive understandings of racial worth on Panamanians. The anger felt by the Panamanians as a result of such encounters was intensified by fears of annexation and filibustering. The US had recently taken a large chunk of

Mexican territory and US filibuster William Walker was flexing his muscles in Nicaragua. It was in this context, according to McGuinness, that the term "América Latina" was coined as a means to defend Panamanian sovereignty and denounce US actions. Panamanian politician Justo Arosemena made a distinction between the Latin races, who he believed were less materialistic and more spiritual, and the Yankees, who had aggressive and conquering desires.

So we need to question what Mignolo (2005: x) refers to as "an excess of confidence . . . regarding the ontology of continental divides". The distinct trajectories taken by the areas north and south of the Río Grande are based on colonial and Eurocentric logics; and the divide posits a mode of understanding that excludes indigenous and African perspectives. Mignolo (2005) urges us helpfully to think of Latin America as an invention rather than a discovery, as a starting point for rethinking the continent in a way that includes marginal and decolonizing perspectives. Disrupting ontological and cartographic certainties means recognizing that Latin America does not have, nor does it require, a fixed definition. Indeed, one cannot definitively assert which countries belong to Latin America and which are excluded, despite the attempt to do so in Table 1.1. Indeed, the countries included in or excluded from Latin America as a region depend on territorial, linguistic or political definitions. While the Spanish- and Portuguese-speaking countries of the Central and South American mainland are almost always included, these territories also contain English-speaking (Belize, Guyana, the Atlantic Coast of Nicaragua), Dutch-speaking (Suriname) and French-speaking (French Guiana) territories, which are often excluded. I have included them because of their location on the Latin American mainland.

The nations of the Caribbean also pose definitional complexities. While almost all of them could be included or excluded, I have included Cuba, the Dominican Republic, Haiti and Puerto Rico, because of their identification with Latin America by many of their citizens as well as the multifaceted influence of these nations on Latin America politics and development. In other words, while the inhabitants of these Caribbean nations might inhabit diverse modes of regional identification, which for many would certainly include Latin America as well as other regional identifications (the Caribbean, France, the Black Atlantic), the historical and intellectual influences of Cuba, the Dominican Republic and Haiti on Latin American development and the shared development traits between the trajectories of these three nations and Latin America as a whole (which might include *inter alia* legacies of slavery, commodity trade, dictatorships, revolutionary struggle and US intervention) cannot be dcnicd. In other words, it is important to note that there are competing regionalizations at work, Two of the nations included, French Guiana and Puerto Rico, do not yet have independent status. French Guiana is an overseas region of France, and while it belongs to the European Union and uses the euro as currency, it is economically integrated in South America. Puerto Rico is a Spanish-speaking unincorporated territory of the United States and has a strong political movement pushing for independence. However, there are now more Puerto Ricans living on the US mainland than there are on the island (Ennis et al. 2011).

At the present time, a number of important economic, cultural and political processes are under way which are redefining Latin America's role in the world and its relationship with the United States. First, in the past few years we have witnessed major transformations in the Latin American electoral landscape. The election of populist left-wing leaders, such as Evo Morales, Bolivia's first indigenous president, and Venezuela's Hugo Chávez, shows a decisive rejection of the neoliberal model that has been in place for the past two decades. Second, while the United States continues to intervene in the economic and political life of the region, this intervention is increasingly challenged by investments and trading relations from China and Korea and stronger intra-continental relations within Latin America. Third, Latin America's indigenous and black populations are experiencing a cultural and political resurgence, generating new ways of doing politics and challenging deeply embedded forms of racism and social exclusion. Finally, the growth of Latin American populations in the United States means that US political candidates can no longer ignore the importance of Latino voters. In addition, Latin America's cultural production – its telenovelas, its cinema, its music – has gone global in ways that rework understandings of the continent held by outsiders.

Despite these transformations there are also many continuities, and substantial development challenges lie ahead. Free trade and privatization deals continue to be signed, which often exacerbate the existing vulnerabilities of marginalized populations. There is evidence that democratization remains fragile in some places. For example, Honduras experienced a coup in 2009 that removed the democratically elected president from power, and since then human rights abuses which include the assassination of journalists and activists have escalated. Mexico is currently engaged in a bloody and escalating drugs war. Violent gangs are active in a number of urban spaces in Central America. While poverty has decreased for some Latin Americans, inequalities between rich and poor continue to be resilient. The struggle for sustainable development in the

Amazon, in the rainforests of Central America, in the Andes and in the coastal lowlands faces major and substantial obstacles as the inhabitants of Latin America deal with the effects of climate change and other major disasters such as earthquakes, hurricanes, droughts and floods.

When we think of Latin America and Latin Americans, it is therefore important to gain an understanding of the region's cultural and geographical diversity and specificity, but to do so without exoticization of Latin America's problems and challenges. Much of the fascination felt towards Latin America by outsiders is unfortunately frequently accompanied by simplification and stereotyping. As Skidmore and Smith (2005: 6) point out, Latin America and its inhabitants have been subject to "a jumble of racist epithets, psychological simplifications, geographical platitudes and cultural distortions" by outsiders. From outside the continent, Latin America has been widely admired but as the first part of the world to be colonized by European powers, for many outsiders it appears to embody the quintessential third world region (see Black 2011), subject to the exoticizing gaze of the first world along with its stereotypical tropes. Like other parts of the so-called third world, Latin America has frequently been seen as poor and excessively prone to political instability, corruption, disease, natural disasters as well as unbridled passions and unrestrained machismo. While neither machismo nor social decomposition is absent, such features should not come to define the continent as a whole. Latin America constantly defies these stereotypes, providing examples of optimism, civilization, political mobilization and resistance, participatory democracy, creative forms of cultural expression, and a willingness to innovate, experiment and take risks. A starting point is to recognize that Latin Americans are very similar to US Americans and Europeans in many respects; they go to school (see Figure 1.5) and university, run businesses, download music, text friends, have parties, get married, get involved in political, economic and environmental causes, and they both embrace and resist globalization. On the other hand, there are important cultural and political differences of which a student or practitioner of Latin American development must be cognizant in order to avoid making *a priori* and problematic assumptions about Latin American realities, needs and desires.

While different disciplines and studies might emphasize the political, the economic, the social, the cultural or the environmental aspects of Latin American development, in practice these dimensions can only be separated for analytical purposes and gaining an in-depth understanding of contemporary Latin American development means holding all of them in tension. Forms of economic exclusion have political effects, and there

Figure 1.5 *While many Latin American children are unable to attend school, it is important to recognize that for most children attending school is a normal and compulsory part of life. Schoolchildren in Matagalpa, Nicaragua (top) and Santiago, Chile (bottom)*

Sources: Julie Cupples (top photo), Natasha Vine (bottom photo).

is a cultural politics to the opening and closing of economic spaces. Gaining access to the fruits of development (education, clean water, gender equality) is never merely about access to material resources; it is always about identity, meaning and asserting the right to exist with dignity and pride (see Moore 1996).

Latin American development

It is impossible to gain a sense of Latin American development without some broader understanding of development theories. Just as the "Latin" in Latin America needs to be unpacked and interrogated, what we mean by "development" in the phrase "Latin American development" demands the same level of enquiry. Development theory can be a confusing mass of -isms for the uninitiated, informed as it is by modernism, liberalism, capitalism, socialism, Marxism, neoliberalism, postcolonialism, poststructuralism, feminism and environmentalism. Gaining some sense of these -isms, their origins and how they have been used over time both within and beyond Latin America is important.

It is also important to consider theories of development endogenous to Latin America. There is a tendency to think of development as something done to Latin America by more powerful first world nations, but Latin American intellectuals have been theorizing about the question of development since the colonial era and it is valuable to assess which theoretical paradigms and approaches might have explanatory power for the present moment and for research into specific development issues. At the same time, we need to recognize development as something that is part of everyday life for many Latin Americans, who might variously be seen as the compliant, resistant or ambivalent targets or subjects of development. In other words, it is important to think of development not only as a scholarly or professional endeavour but also as something practised, embodied and performed into being by ordinary people in the spaces of everyday life.

In broad terms, development is about generating greater well-being and standards of living. It explicitly entails a need and desire to bring about change to a situation that is deemed by some to be unacceptable or undesirable. What is to be done to address this state of affairs can never be objectively determined. To be sure, we all need food, shelter and clean water, but how these things are obtained, delivered and used is subject to individual and collective interests and underpinned by ideologies and relations of power. In the first couple of decades after the Second World War, development was viewed largely as a technical issue, a solution to an obvious problem or lack, devoid of and separate from ideology. Today, however, it is apparent that such an approach is flawed, and many scholars understand development to be a politically and culturally mediated process. All those involved in the development process – development practitioners, beneficiaries, scholars and activists, the agents and subjects of development – make sense of development in cultural ways, drawing on dominant and alternative discourses, life experiences and political ideologies. Of course, many of us occupy multiple and shifting subject positions with respect to the development project, and reflecting on our subject positions, as well as those occupied by others, is central to gaining a more sophisticated understanding of development processes and dynamics.

It was in the postwar period that modernization theory began to take hold in development circles. Indeed, Colombian anthropologist Arturo Escobar (1995) believes that it was in the period after the Second World War that the idea of development as we currently understand it was brought into being. At this time, the United States, less affected by the Second World War, was enjoying economic ascendancy, a number of nations in Asia and the Pacific were undergoing decolonization and new kinds of global relationships were being forged. It was also a time of optimism about the prospects for economic and technological progress, and the United States was anxious to gain access to emerging markets. According to Escobar (1995), the inaugural address of US President Harry Truman in 1949 can be seen as a landmark moment that heralded the contemporary development age. Truman asserted:

> More than half the people of the world are living in conditions approaching misery. Their food is inadequate, they are victims of disease. Their economic life is primitive and stagnant. Their poverty is a handicap and a threat both to them and to more prosperous areas. For the first time in history, humanity possesses the knowledge and the skill to relieve the suffering of these people . . . I believe that we should make available to peace-loving peoples the benefits of our store of technical knowledge in order to help them realize their aspirations for a better life . . . What we envisage is a program of development based on the concepts of democratic fair-dealing. Greater production is the key to prosperity and peace. And the key to greater production is a wider and more vigorous application of modern scientific and technological knowledge.
>
> (US President Harry Truman, 20 January 1949,
> cited in Escobar 1995: 3)

So Truman's approach amounted to defining half the world in negative terms, as suffering, poor and underdeveloped and the other half, the western world, as in command of the knowledges and technologies to address this suffering (Escobar 1995). This mode of thinking took hold in the first world, and had serious consequences for the parts of the world deemed to belong to the third world, namely Africa, Asia, Latin America and the Caribbean. The idea that with western help, knowledge and technologies third world countries would modernize and "catch up" with the industrialized countries became pervasive and gave rise to the development industry we know today, based on multilateral and government aid agencies, non-governmental organizations (NGOs) and experts (economists, agronomists, demographers, geographers, sociologists and so on) based in universities or think tanks. It was assumed that indigenous and non-European cultural traditions would be abandoned as people adopted modern values and practices.

The invention of development (Escobar 1995) at this time was not purely about economic motives; there were also strong geopolitical motivations. This was the start of the Cold War and concerns were developing within the US administration about the spread of communism and the appeal of communist ideas to disenfranchised and impoverished groups of people in the third world. One of the most influential contributions to modernization thinking was Walt Rostow's *The Stages of Economic Growth: A Non-Communist Manifesto*, published in 1960: the geopolitical underpinnings of modernization theory are clearly depicted in the book's subtitle. Rostow (1916–2003) advised on national security during the presidencies of John F. Kennedy and Lyndon B. Johnson and was a committed free market capitalist and anti-communist. Capturing the hearts and minds (and economies) of the third world and integrating these economies into a global capitalist system was viewed as an important anti-communist strategy, which would prevent third world countries turning to socialist or communist thought and ideals. Rostow's influential thesis posited that the countries of the third world were much like the pre-industrial communities of Europe and therefore they had to experience the conditions of economic transformation and social change that European countries had already experienced during the industrial revolution. Indeed, for Rostow, modernization comprised five distinct stages: traditional society, preconditions for take-off, take-off, drive to maturity and high mass consumption. Development was thus seen as a linear process towards the level and type of development found in the West. The big difference between developed and developing countries was that the developing countries had to be *helped* on their path to development by the industrialized nations, while the first world had developed by its own means. And so to become developed meant adopting first world practices, ideas, knowledges and investment.

This is a valuable perspective in terms of interrogating the terms on which development is evoked and enacted. It is, however, important to recognize that while development might have been "invented" by the first world, once it is in discursive circulation in Latin America and other areas designated "third world", it is immediately subject to re-articulation by both intellectuals and ordinary people. Rostow's and Truman's ideas, while powerful, could not be definitively stabilized. Indeed, not long after modernization theory had taken hold of expert thinking, Latin American and other intellectuals began to develop an alternative theory of development, which coalesced in what came to be known as dependency theory. Dependency theory questioned the key assumptions of modernization theory by arguing that Latin America could never follow the linear trajectory to development outlined by modernization theorists because the development of the industrialized world had occurred at Latin America's expense. If Latin America was to embark upon a successful form of development, it needed weaker rather than stronger ties with the industrialized nations and greater self-sufficiency.

Development as a discourse

One of the key theoretical points to understand about development is that it gets mobilized discursively. Discourses are modes of understanding and talking about an issue that are embedded in relations of power. Discourses create shared understanding and are central to the ways in which we make sense of the world. Discourses about places or people in Latin America, particularly those put into circulation by dominant social groups, have material outcomes and shape policy interventions and media representations. But marginal and subordinated populations are capable of generating alternative discourses which constantly pose a challenge to the stability or legitimacy of dominant discourses. To say that development is a discourse therefore is not to say that development is not real, or does not have material manifestations, but to show that how we think, talk and write about development has a bearing on how we "do" development. Escobar (1995) has shown how development discourse is not an objective representation of development, but rather brings development into being. He has also shown how the discourse of development that has been a dominant feature of the postwar period has had insidious outcomes for the targets of development, principally those deemed to be poor or lacking in Africa, Asia and Latin America. He observes that development as a discourse has been so pervasive and persuasive that inhabitants in these parts of the world have internalized feelings of inferiority. They have come to see themselves as needing "development".

Discourses, in a Foucauldian or poststructuralist sense, are always about power and they are always unstable. Despite their inherent instability, some discourses pertaining to development and the idea of the third world can be very resilient. So to evoke development is a tricky business because it always means asserting some kind of power relationship, either about how the world is or about how the world should be. What constitutes good development for one group or individual might be rejected as inappropriate by another. The instability of discourse means, however, that it is possible for alternative ways of both understanding and doing development to gain traction. A substantial number of development theorists from Latin America and beyond, such as Escobar and Gustavo Esteva, have drawn our attention to development's failures, pointing out how in many respects the development project has been a dismal failure because in spite of 50 years of development practice, research and aid, people still lack basic necessities – dying of malnutrition or easily preventable diseases, failing to complete primary or secondary education, living with housing or income insecurity – an approach that has come to be known as postdevelopment (see Esteva 1987; Edwards 1989; Escobar 1995; Rahnema 1997).

Thinking about development's failures and thinking in terms of postdevelopment does not mean rejecting development, because the possibility of re-articulation always exists. As Cupples, Glynn and Larios (2007) have demonstrated, drawing on the insights of Italian theorist Antonio Gramsci, even if development has produced some negative and insidious outcomes in countries that dominant and Eurocentric geopolitical imaginaries have designated as third world, the subjects or agents of development can never be definitively fixed into a set of understandings. The material conditions of everyday life that many Latin Americans face – hunger, inadequate access to health care, underemployment, flimsy housing, an inability to complete school – act to disrupt the common sense, replacing it with a "good sense". In other words, people do have a sense of what they're up against and do try to create a better and more dignified life for themselves. They can *re-articulate* the dominant discourses of development, to come up with something more empowering and more inclusive. Moreover, this struggle can only take place within the existing ideological terrain (Cupples *et al.* 2007). It would therefore be pointless to turn up in a low-income rural or urban community in Latin America and inform people who are mobilized to improve their standard of living that development was a big con and they need to overthrow it, because this group of people will already be in the process of taking ownership of development, re-articulating the concept in a way that makes sense to them.

Gramsci used the concept of hegemony to understand the nature of political struggle and urged us to think in terms of processes rather than final once-and-for-all victories. This process involves recognition that contestation over the dynamics of development will necessarily be ongoing and the struggle over meaning will never be complete (see Hall 1996; Condit 1994; Cupples *et al.* 2007). This is a particularly useful approach for both studying and practising development in Latin America. As experienced development practitioners know, "just as one problem is attenuated, others arise; as soon as one group of people finds a measure of satisfaction, others begin to express grievances" (Cupples and Glynn 2013: 10). Development does not therefore have an end point and is continuous.

In a recent book dealing with development in Central America, Wainwright (2008) draws on the work of Indian postcolonial feminist Gayatri Spivak and asserts with a double negative that "we cannot not desire development". Wainwright shows how criticizing, deconstructing and interrogating development is necessary but, in spite of its flaws, doing away with development is just not possible in the current conjuncture. In other words, development might have emerged in the context of colonizing and modernizing processes, it might be a neocolonial technique of power in many ways that has indeed had negative, disempowering and insidious outcomes for many, but we still have a moral responsibility to respond to conditions of suffering in the world.

There is still much suffering in Latin America that demands our attention, as scholars, activists, practitioners and citizens. Latin America is the site of some of the worst forms of human rights abuses and exploitation the world has ever known. These abuses occurred both in the sixteenth century and in the 1970s and 1980s. The levels of cruelty and brutality exacted on those forced to work in mines or plantations to generate wealth for the colonizers, those deprived of their land and unable to feed their children, and those deemed subversives by their own government and disappeared, tortured and murdered are chilling reminders of the distortion of humanity served up by history. Violence, poverty and exclusion still underpin the frequent murders that take place in Ciudad Juárez on the US–Mexican border, the mobilization of indigenous peoples in political organizations and guerrilla armies, and the desperate dash across the US–Mexican border in order to become an illegal, undocumented and discriminated migrant in the United States. At the same time, Latin America is also the site of the most extraordinary and heroic forms of courage and resistance to these same forces. The ways in which so many ordinary men and women have stood up to such brutality in defence of dignity, love and hope reveal Latin America's enduring humanity. Latin America can be so depressing but is often so inspirational.

Diverse geographies

Latin America covers a large geographic area from the US–Mexican border in the north to the southern shores of South America, and often includes a number of Caribbean islands. It is composed of more than 20 nation states, a land mass of 21 million square kilometres and a population of almost 600 million, 40 million of whom are considered or consider themselves to be indigenous. It is frequently broken up into sub-regions, such as (part of) North America, Central America, South America and the Caribbean (see Figure 1.6).

Most Latin Americans speak Spanish, but Portuguese is spoken in Brazil, the largest country in Latin America. Many indigenous languages are also

Figure 1.6 *Map of Latin America*

Source: Marney Brosnan.

spoken, including Quechua and Aymara in the Andean region and more than 20 Mayan languages in Guatemala. Mexico alone has more than 200 different languages.

Latin America is far more linguistically diverse than Europe. While Europe has only two language families and one isolate language (a language that is not related to other languages), Latin America has more than 50 language families and more than 70 isolates. While hundreds of different languages are still spoken, many of these are seriously endangered. Since colonialism, many indigenous languages have been lost and continue to be lost. On the other hand, there are still a number of vibrant and stable indigenous languages with more than a million speakers, including Quechua, Guaraní, Kekchí and Nahua. Quechua still has more than 8 million speakers spread across Peru, Brazil, Bolivia, Ecuador and Colombia (for more detail see Kaufmann 1994a, 1994b and www.ailla.utexas.org).

To describe Latin America's population as diverse is an understatement. Latin America's population is composed predominantly of European, African and native American heritages and a large variety of mixtures thereof. While Argentina and Uruguay see themselves as very European nations, most Bolivians and Guatemalans identify as indigenous, while Cuba, Colombia and Brazil have large black populations, descended from African slaves. Other countries, such as Mexico and Nicaragua, see their populations and their cultures as predominantly *mestizo* or mixed. The continent is home to former British and Dutch colonies, such as Belize, Guyana and Suriname, with very different cultural and linguistic heritages, and Jews, Syrians, Italians and Poles have also settled in Latin America (Wade 1997). These dominant cultural categories are extremely fluid and at times heavily contested. In some parts of the continent, processes of native re-identification or of newly asserting a previously marginalized or invisible black or Afro-Latino identity are under way.

While Latin America is often known for its natural environments (the Andes mountains, the Amazon rainforest), the Latin American population is overwhelmingly urban, with more than 75 per cent of the population living in cities (González 2011). Many of Latin America's cities are extremely large by global standards – the populations of Mexico City and São Paulo exceed 20 million inhabitants, Buenos Aires has 13 million, while Rio de Janeiro, Lima, Santiago and Bogotá all have in excess of 5 million inhabitants. While the vast majority of Latin Americans consider themselves to be Catholics, the religious landscape is undergoing dramatic transformation as a result of both secularization and the growth of Protestant and especially US Evangelical churches in the region.

As Latin America constitutes a large land mass, it has a diverse range of climates and environments. Much of the region is tropical but it also has some mid-latitude territories and a number of outstanding biophysical features, which include the world's highest settlement, airport, volcano, railroad and highway, the world's second largest river and highest waterfall, as well as the largest tropical rainforest in the world (González 2011). It is believed that the Amazon rainforest produces around 40 per cent of the world's oxygen supply (Wiarda and Kline 2011). The range of biodiversity present in Latin America is staggering and includes crocodiles, jaguars, tapirs, spider monkeys, toucans, quetzals, turtles, boas, anacondas, hummingbirds, flamingos, llamas and alpacas. Some of these birds and mammals are seriously endangered.

In Latin America one can find snowy mountains, dry deserts, lush highlands and fertile coastal lowlands. González (2011) has identified 13 different natural regions, all with their own unique characteristics and economic and agricultural activities. They include the Amazon basin, the Orinoco basin, the Antilles archipelago, the coastal lowlands of Mexico and Central America, a highland cordillera stretching from Mexico to Bolivia, the Brazilian highlands, the pampas of Northwest Argentina and the Atacama desert region of Peru (see Figures 1.7–1.9). These diverse landscapes have facilitated a range of economic and agricultural

Figure 1.7 *Torres del Paine, Patagonia, Chile*

Source: Marney Brosnan.

activities. Latin America has produced and produces maize, beans, rice, potatoes, rubber, coffee, sugar, cacao, wheat, bananas, citrus fruit, mangoes and flowers; it has mined gold, silver, copper and tin; herders and ranchers have reared llamas, alpacas, sheep, goats and cattle, while fishing communities fish for shrimp, lobster and red snapper. In addition, Latin America's natural landscapes attract large numbers of tourists and visitors who travel to the region to experience its forests, mountains and wildlife. Much of the continent lies on an earthquake fault line or in a hurricane belt and the combination of those biophysical characteristics with weak socio-economic development and preparedness means Latin America has also been home to some of the world's most devastating disasters.

Figure 1.8 *Orinoco River, Amazonas State, Venezuela*

Source: Pedro Gutiérrez, Wikimedia Commons
(Creative Commons Attribution 2.0 Generic).

Figure 1.9 *San Pedro de Atacama, Atacama Desert, Chile*

Source: Entropy 1963, Wikimedia Commons (released into the public domain).

Outline

This book surveys all the principal facets and dimensions of Latin American development. It is grounded in critical contemporary approaches to development, informed primarily by Marxist, poststructuralist, postcolonial and feminist theory, but it also draws readers' attention to dominant liberal and neoliberal approaches to the question of development. Chapter 2 will provide an introduction to Latin American history, from the pre-Columbian era to the start of the twentieth century. Chapter 3 focuses on twentieth- and twenty-first-century economic history and processes, and in particular on the rise of the neoliberal economic model, its impacts and the ways in which it has been challenged. Chapter 4 looks at Latin America's political transitions and transformations since the beginning of the twentieth century and covers Latin America's relationship with the United States, its dictatorships and revolutionary movements and its more recent political developments. Chapter 5 focuses on Latin America's environments, emphasizing the multiplicity of factors that have worked against environmental sustainability. Chapter 6 focuses on identity politics and how the dynamics of race, gender and sexuality shape development processes. Chapter 7 is devoted to the region's indigenous peoples, and explores their struggle for survival and the ways in which they have mobilized in defence of a more indigenized mode of development. Chapter 8 explores the role of media and popular culture in enacting and contesting development. The book concludes with a chapter that reflects on the current situation and the extent to which processes of decolonization are under way in Latin America.

Summary

- Studying Latin American development involves grappling with concepts such as globalization and hybridity.

- Latin America defies easy generalization and does not have a fixed definition.

- Latin America is a site of both wealth and poverty.

- A number of different theoretical approaches are used in the study of Latin American development.

- Development as we understand it today took hold after the Second World War.

- It is valuable to conceptualize development as a discourse.

- Latin America is geographically diverse.

Discussion questions

1. Why could it be considered problematic to say "America" when referring exclusively to the United States?

2. Study the statistics in Table 1.1. Discuss the ways in which such development indicators are useful and ways in which they might be considered limited or problematic.

3. What is meant by the concept globalization, and how does it relate to Latin American development?

4. What are the main theoretical approaches used for studying Latin American development and what do they emphasize?

Further reading

Escobar, A. (1995) *Encountering Development: The Making and Unmaking of the Third World*. Princeton, NJ: Princeton University Press. Written by a Colombian anthropologist whose work has been central to Latin American studies and development studies, this is an important book that reflects on how development was brought into being in the postwar era and how it functions discursively.

Galeano, E. (1973) *The Open Veins of Latin America: Five Centuries of the Pillage of a Continent*, trans. C. Belfrage. New York: Monthly Review Press. Originally written in 1971 in Spanish by the Uruguayan author, this book has become a classic and an absolute must-read for anybody interested in Latin American development.

Green, D. (2006) *Faces of Latin America*, 3rd edn. London: Latin America Bureau. An engaging and informative basic introduction to Latin America.

McEwan, C. (2009) *Postcolonialism and Development*. London: Routledge. Focusing on the relationship between postcolonialism and development, this book contains some accessible introductions to development theory.

Mignolo, W. D. (2005) *The Idea of Latin America*. Malden, MA: Blackwell. An intellectually stimulating critique of the idea of "Latin" America.

Willis, K. (2005) *Theories and Practices of Development*. London: Routledge. A good introduction to theories of development.

Websites

www.ailla.utexas.org/site/welcome.html
> The Archive of the Indigenous Languages of Latin America (AILLA). Based at the University of Texas, a website containing information about the indigenous languages of Latin America.

http://news.bbc.co.uk/2/hi/country_profiles/default.stm
> The BBC has detailed profiles on all of the territories and countries in the world, with data on history, politics, economics, leader, media and links to BBC audio and video archives.

http://lanic.utexas.edu
> Website of the Latin American Network Information Center (LANIC) at the University of Texas at Austin. LANIC has a wealth of links to information in and about Latin America. An indispensable resource for students, academics and professionals.

www.guardian.co.uk/global/interactive/2009/apr/18/country-profiles-world-map
> Use *The Guardian* newspaper's interactive world map to link to news stories on Latin America, organized according to country.

www.coha.org
> The Council on Hemispheric Affairs is a private, nonprofit research organization that produces regular short articles on Latin American affairs and US–Latin American relations.

http://upsidedownworld.org/main
> *Upside Down World* is an online magazine which contains independent articles, news briefs and blogs on Latin American politics development aimed at interrogating corporate globalization in the region and focused on grassroots and people-centred initiatives.

2 A brief history of Latin America

Learning outcomes

By the end of this chapter, the reader should:

- Have developed a basic understanding of the conquest, colonization and independence of what is now called Latin America
- Appreciate the ways in which indigenous cultures resisted colonization
- Be familiar with key colonial legacies such as commodity trade

The creation of Latin America

Some understanding of Latin American history is fundamental to gaining insight into how Latin America as a region came to be so intimately bound up with the development project of the twentieth century. Indeed, many of Latin America's contemporary development challenges as well as the diverse ways in which development is accommodated and contested are embedded in historically generated relations of power. The past provides ongoing obstacles to contemporary development as well as inspiration on how development might be better enacted.

The region we now call Latin America has been settled for many millennia. The first human migrants to cross the Bering land bridge did so around 40,000 BCE. Before European conquest and settlement, this region was home to many large, complex and diverse civilizations, including the Zapotec, Mixtec, Olmec, Toltec, Mexica (Aztec) and Mayan cultures of Mexico and Mesoamerica, the Valdivia culture of Ecuador, the Chavín, Moche and Inca cultures of Peru, and the San Agustín and Muisica cultures of Colombia. In the pre-Columbian era, there were nomadic hunter-gatherer tribes as well as peoples who practised agriculture, cultivating crops such as maize, manioc, squash, potato, chillies and cotton. Many of these indigenous cultures had advanced knowledge systems. The Mayans had a sophisticated writing and calendar system, which linked to agricultural, cosmological, astronomical, religious and mathematical knowledges. The Aztecs were skilled warriors and had advanced architecture and modes of urban planning. The Incas had well-developed engineering skills in transport systems and irrigation for agriculture. The pre-Columbian period also included many practices that would not be considered civilized or humane by today's standards. Aztec Mexico for example frequently saw human sacrifice including that of children, and the emperor Montezuma reportedly kept a human zoo populated by dwarves, albinos and hunchbacks (de Sagahún 1829). The cultural, artistic, linguistic and religious practices of the pre-Columbian peoples have fascinated generations of scholars and they continue to be significant in the present in all kinds of ways.

Some of the indigenous civilizations that existed prior to the Spanish conquest collapsed for reasons that are poorly understood, although climate change, drought, famine, political rebellion and war are likely contributing factors. Others, such as the Aztec empire of Mexico and the Inca empire of Peru, flourished and became mighty empires. It is important to acknowledge the heterogeneity of the indigenous population and to recognize that pre-Columbian Latin America comprises complex histories that defy easy categorization. As Williamson (2009: 84) notes, the term "Indian" is used to distinguish native Americans from Europeans but "it should not be taken to denote some uniform seamless culture". While it is difficult to do, students of Latin American development should avoid viewing these histories through presentist lenses, as the values and discourses through which we make sense of development today would not have had any functionality in these times and places. One should avoid romanticizing the pre-Columbian past as a time of great ecological harmony or exoticizing it as a time of bizarre ritual cruelty. It is, however, necessary to think about the ways in which the pre-colonial and colonial past continues to shape contemporary Latin American development, as contemporary development dynamics are shaped by indigenous cultures with much longer continental histories. At the same time, it is valuable to reflect on the legacies of the colonial past for the present and adopt what might be referred to as a postcolonial perspective. The prefix "post" in postcolonialism is not one of chronological temporality: it does not simply refer to the period after colonialism but rather the ways in which contemporary cultures, social practices and power relations continue to be influenced by the colonial past (McEwan 2009).

The rise and fall of indigenous civilizations, the practice of war and ritual sacrifice, and exposure to environmental and social calamities mean that the pre-conquest peoples of the continent endured a tumultuous history. It is possible however that such historical experiences did little to prepare native Latin Americans for the violence, brutality and cultural conflicts that would ensue from the fifteenth century onwards as a result of the arrival of Europeans.

The conquest of Latin America, and thus its emergence as a geographical entity, began in August 1492 when Christopher Columbus, with the financial support of the Spanish king and queen, set sail from Spain, making landfall in the Caribbean on 12 October. This and other voyages of discovery had a range of motivations. Given the intense imperial competition over trading routes between European powers, Columbus was determined to find an alternative route to Asia, primarily India, China and Japan, which would strengthen Spain's position within Europe. In addition, the first conquistadores were driven by a desire to establish profitable trading posts, from where precious metals and agricultural products could be shipped back to Spain. And finally, the colonization of America was seen as a means to promote and spread Catholicism. It is important to recognize that Spain's colonization of the Americas came after 700 years of Muslim occupation. The struggle between Christianity and Islam became so deeply embedded in the Spanish psyche that it was as important to convert the native peoples to Christianity as it was to convert them into slave labour in mines and on plantations in order to generate wealth for Spain. As Bakewell (2011: 80) writes, the Spanish colonizers "freely admitted that they had conquered America for the sake of God and gold". In other words, it was as much a spiritual conquest as a political, military and economic one.

Columbus made a total of four voyages to Latin America but died believing he had reached Asia as was intended. Consequently, he referred to the native inhabitants as *indios* (Indians), a label that has stuck, although not without contestation and re-articulation. It was not until Amerigo Vespucci, from whom the name America is derived, participated in a series of Portuguese voyages to colonize what is now Brazil that it became recognized that a new continental land mass had been discovered. Subsequently, Latin America began to be referred to by Europeans as a New World. For much of the colonial period, Spain referred to Latin America as the Indias Occidentales. It was not until much later than the terms America and Latin America became part of dominant currency.

The colonization of Latin America began in the Caribbean and spread north to Mexico and Central America and later south towards Peru. Spain's initial failure to recognize that a new continent had been discovered is in part what led to the Portuguese colonization of Brazil. Spain conceded this part of the continent to Portugal through a diplomatic treaty, known as the Treaty of Tordesillas, without apparently realizing the significance and expanse of what would later be granted.

While Portugal controlled Brazil, Spanish America was an extensive territory and stretched from the southern point of South America into areas that today make up part of the United States, and it also included a number of Caribbean islands (Bakewell 2011).

The conquest of Latin America was violent and brutal. Many Spaniards had little respect for the indigenous inhabitants, stealing the most fertile land and converting the indigenous population into an exploited labour force. The conquest of the continent has been correctly characterized as genocide as native populations were killed by overwork and ill-treatment in mines and on plantations and by "Old World" diseases such as influenza, smallpox and measles, against which they had no immunity.

The size of the native population in 1492 is a matter of controversy and debate, and historians have produced widely varying estimates (Denevan 1992a). Denevan (1992b) estimates the total population of the Americans in 1492 to be around 57 million – 4.4 million in North America, 21 million in Mexico, 5.7 million in Central America, 5.9 million in the Caribbean, 11.5 million in the Andes and 8.5 million in lowland South America. Skidmore and Smith (2005) believe that the population of Mexico declined by as much as 95 per cent between 1519 and 1605, while Schwerin (2011) states that the indigenous population of the Caribbean was wiped out in only 50 years.

When the Spaniards began to conquer a new territory, the indigenous inhabitants were read a document known as the *Requerimiento* which informed those that stayed to listen that they were to "lay down their arms, submit to the king of Spain, and embrace the Catholic religion. If not, the conquerors would not be held responsible for the consequences" (Merrell 2004: 54). Colonization proceeded on the basis of a system known as the *encomienda* which granted native workers to Spanish settlers on the understanding that they would care for them. Most *encomenderos* continued however to exploit and extort the native people, through overwork, the imposition of taxes and forced religious donations. As a result, they often earned the horrified condemnation of the missionaries who had been sent to the Americas to engage in evangelization work (Williamson 2009). Some of these, such as Friar Bartolomé de las Casas, the first Bishop of Chiapas, began to defend the human rights of the Indians and document the atrocities committed by the *encomenderos*. Las Casas did however continue to believe that evangelization was in their interests.

Williamson (2009) believes that it was never really resolved in the colonial period whether Indians were free or enslaved, but there is no doubt that they were subject to horrendous forms of exploitation. The high native death toll combined with a voracious demand for labour meant that millions of slaves were brought to the Americas from Africa to make up for the labour shortfall, with the first African slaves making the journey in 1518.

Government of the new colonies was a logistical challenge, given the communication and transportation capabilities of the period and the size of the area to be governed. To control its territories, the Spanish crown created two viceroyalties, the Viceroyalty of New Spain in 1535 and the Viceroyalty of Peru in 1544. Viceroyalties were administrative territories ruled by viceroys based in Latin America rather than in Spain, to whom authority was delegated. The viceroy of New Spain was based in Mexico City and his jurisdiction included Mexico, much of Central America and the Caribbean colonies. The viceroy of Peru controlled Panama and all of the Spanish territories in South America. In the 1700s, two further viceroyalties, New Granada and River Plate, were created to administer the areas that today make up Colombia and Argentina respectively.

The viceroyalties were administratively divided into provinces controlled by a governor and subject to a court of law, known as an *audiencia*. Other political leaders at provincial level included the *corregidores* whose role was to collect taxes, apply justice and organize Indian labour. The Catholic Church was also a powerful institution and bishops frequently intervened in political matters. Colonial Latin America was a hierarchical bureaucracy in which corruption was common and that provoked widespread resentment and resistance from native peoples subject to its logics and its jurisdiction.

After Columbus, many Spaniards travelled to Latin America in search of wealth and fame, often inspired by mythical tales, such as that of El Dorado, which promised great prosperity should it be found. The colonization of the continent was at times patchy and erratic. Many areas were barely settled and of little interest to colonizers because of a lack of mineral resources, while others were subject to destruction and dramatic transformation. In some places the indigenous populations were wiped out; in others they continued to survive and maintain their traditional ways of life. One of the most well-known conquistadores was Hernán Cortés, who moved into the great Aztec capital of Tenochtitlán, the site of present-day Mexico City (see Box 2.1).

Box 2.1

Cortés in Mexico: laying waste to Tenochtitlán

When Hernán Cortés advanced on Tenochtitlán, it was the seat of a vast empire ruled by the Aztec emperor, Montezuma. Tenochtitlán was an impressive feat of civil engineering and urban planning. Built on a swamp, it possessed awe-inspiring temples, ceremonial buildings, gardens, markets, sports arenas, zoos and sacrificial platforms. It had wide streets, avenues and a series of interconnecting canals adorned with *chinampas*, floating gardens in which crops and flowers were grown. The city's inhabitants enjoyed fresh water from mountain springs and the city deployed an efficient sewerage and garbage collection system. The nobles lived in opulent luxury and abundance. Cortés and his men were overwhelmed by Tenochtitlán's beauty and magnificence. Cortés described it as the most beautiful city in the world and then he ordered his men to lay the entire city to waste.

To defeat the Aztecs, Cortés formed a propitious military alliance with the Tlaxcalans, their fervent enemies. Despite the advantages to Spain accrued through such an alliance, many historians are still baffled by the apparent ease with which Cortés was able to defeat the Aztec empire (Williamson 2009). It is possible that his victory was facilitated by an Aztec belief that he was the embodiment of a Mesoamerican deity, Quetzalcoatl, but that view is no by means uncontested (Restall 2004; Skidmore and Smith 2005). Cultural difference is likely to have played an important role. While the Aztecs had substantial military prowess, they went to war for very different reasons and did not seek to destroy enemy territories. The Spaniards also suffered heavy casualties in their battles to conquer the Aztecs. In one night in 1520, 1000 Tlaxcalans and 900 Spaniards were killed.

During the conquest of Mexico, a local woman, known variously as Marina, Malintzin or Malinche, became Cortés' mistress and interpreter. She bore his child, considered to be the first Mexican or mestizo (person of mixed race). Mexican author Octavio Paz (1950) sees Malinche's rape by Cortés and Mexico's subsequent colonization as central to Mexican national identity and the key source of shame and sense of inferiority felt by Mexicans. The last Aztec king, Cuauhtémoc, ruled until 1525. The capital of Mexico, Mexico City, is built on the site of Tenochtitlán.

Just a few years later, conquistadores Francisco Pizarro and Diego de Almagro embarked on similar campaigns to conquer the Inca empire of Peru. At the time, the Inca empire was bitterly divided as a result of a succession war that came about when the emperor and his heir succumbed to an Old World disease, but it still possessed formidable and well-organized armies. However, Pizarro and Almagro took advantage of these divisions and used the brutal massacre of thousands of Indians to further weaken the Inca state.

The new Spanish and Portuguese settlements were turbulent and unruly entities, marked by fierce indigenous resistance to colonial rule, colonial mismanagement and conflict between Spanish settlers, factors that

Box 2.2

The people of maize: the Mayan world

Another great pre-Columbian civilization that survived conquest and colonization is the Mayan culture, whose territory stretches from central Mexico into the Yucatán Peninsula, a small portion of Honduras, the fertile highlands of Guatemala and part of El Salvador. The Mayans built large and complex cities, with brightly coloured pyramid-temples, the ruins of which can be seen today at Chichén Itzá, Tikal and Copán. They practised agriculture, growing corn, cotton, tomatoes, black beans, cacao and sweet potatoes, and hunted animals for food. Mayan beliefs assert that Mayas are people of corn. Corn is still considered to be a sacred crop as well as a staple food in Mexico and Central America today.

The Mayans were ruled by priest leaders and they believed in both corn gods and rain gods. In the face of disaster or calamity, they practised human sacrifice. At Chichén Itzá, victims were thrown into a sacrificial well and it was an honour to be chosen. The Mayans had developed mathematical and astronomical knowledges and they operationalized detailed and complex understandings of time.

The Mayans fiercely resisted the Spanish invasion. In 1517 in the Yucatán Peninsula, they attacked a Spanish expedition led by conquistador Francisco Hernández de Córdoba, who died as a result of the wounds inflicted. Much of what we know about the ancient Mayans today, we have learned from the *Popul Vuh*, or the Book of the People, a K'iche' text, which contains the Mayan creation story and other important cosmological and genealogical narratives. The book was transcribed in both Spanish and K'iche' by Dominican friar Francisco Ximenez in the early eighteenth century. Other important early texts on the Mayans include *The Discovery and Conquest of Mexico*, written by conquistador Bernal Díaz de Castillo (published posthumously in 1632) and the work of Spanish bishop Diego de Landa's *Relación De Las Cosas De Yucatán*, published around 1556. Diego de Landa was responsible for the cruel destruction of much Mayan culture but paradoxically also managed to decipher Mayan hieroglyphics and document these intellectual endeavours.

thwarted and hindered the attempts to establish profitable trading posts. While the great Aztec and Inca empires were conquered by Spain and indigenous populations were decimated, not all indigenous peoples in the Americas were subjugated and neither were their cultures simply obliterated. Some indigenous groups such as the Araucanians in Chile and the Mayans of the Yucatán peninsula proved invincible (Williamson 2009).

Even after Pizarro and Almagro had defeated the Inca empire, they were forced to accommodate and cooperate with Inca rulers and had to battle ongoing Inca resistance. One Inca leader, Manco Inca, ruled Cuzco alongside Spanish rule, initially cooperating with Spanish rulers, but later he rebelled, assembling 10,000 troops and laying siege to the city of

Cuzco. Eventually he fled to the jungles of Vilcabamba, where Inca religious and legal practices were kept alive.

Despite their success in conquering the Inca empire, the lives of both Pizarro and Almagro ended badly. Pizarro, who spent much of his life in jail, was hacked to death by a group of Almagro's supporters, while Almagro was decapitated by one of Pizarro's brothers.

The colonial influence on Latin America is highly apparent. Spain gave Latin America languages, religion, and modes of organizing cities and agricultural production still in evidence today. While many indigenous languages are still spoken, Spanish and Portuguese are the most widely spoken languages on the continent and Catholicism continues to be the dominant religion. All Latin American cities and many towns have impressive Catholic churches built by the Spaniards (see Figure 2.1).

The Spanish also implemented a particular mode of urban planning that is still in place today. Cities were based on a grid system, in which rectangular streets emanated from a central square where the most important buildings were located. The closer one lived to the square, the more important one was (see Figure 2.2). Informal settlements sprang up

Figure 2.1 *The Catholic church in Niquinohomo, Nicaragua*

Source: Kevin Glynn.

Figure 2.2 *A colonial-era house in downtown Buenos Aires, Argentina*

Source: Marney Brosnan.

on the edges of cities, where uprooted indigenous peoples relocated. The Spaniards also created a specific form of unequal land tenure, the *latifundio*, in which large landowners amassed vast quantities of productive land, displacing indigenous peoples and converting them into labour.

The *latifundio* and its counterpart, the *minifundio* or excessively small plot of land on which many plantation labourers and their families relied for basic subsistence, have plagued Latin America's development since the colonial period. The plantation and extractive economy established during the colonial era has been one of the most enduring and most destructive legacies of the colonial era. Mines and plantations have brought social, environmental and economic devastation. The historic centres of gold and silver mining in the colonial period, such as Minas Gerais in Brazil and Potosí in Bolivia, are today sites of backwardness and abject poverty, in part because, paradoxically, the "more a product is desired by the world market, the greater misery it brings to the Latin American peoples whose sacrifice creates it" (Galeano 1971: 61).

Consequently, both native Americans and African slaves were subjected to miserable living conditions while Latin America's wealth financed and fuelled the development of Europe. Gold, silver, nitrates, copper, rubber, indigo were all abundant and heavily demanded by the European colonizers. For a time, after the discovery of considerable silver deposits, the Bolivian town of Potosí was one of the wealthiest places in America, but silver mining proved to be tragic for the native population. According to Galeano (1971), in three centuries a total of 8 million people died working the mines of Cerro Rico in Potosí. Of every 10 people sent to mine silver, seven perished.

The native resources were soon followed by imported ones such as sugar, wheat and coffee, accompanied by similar modes of exploitation. Consequently, Latin America was converted into a supplier of raw materials and its inhabitants into cheap and expendable labour, to fuel the economic development of European nations – a role that to this day it is still attempting to shake off. Moreover, this is a mode of development that did not cease with the formal end of colonial rule. As Mignolo (2005: 55) writes:

> The entire Atlantic economy, from the sixteenth century until the dawn of the twenty-first, was founded on the increasing devaluation of whatever did not sustain capital accumulation. Military defense and political institutions were based on the assumption that human life was expendable in the set of global designs.

Despite the stark inequalities between colonizers and colonized and the violence that underpinned colonialism, it is important to recognize that Spanish and Portuguese culture and customs were never simply imposed on passive native cultures. The colonial period was also a period of heavy cultural borrowing, in which Catholic and pre-Columbian religions and other cultural practices combined to create new forms. What was striking about this period was the ways in which indigenous peoples adopted aspects of Spanish culture and blended them with their own existing cultures (see Box 2.3). The indigenous peoples also developed creative ways of appearing compliant with Spanish rule while simultaneously and surreptitiously mocking, subverting or undermining it. The aims of the Spaniards were constantly subverted by the uses the indigenous people made of the laws, representations and practices imposed on them and from which they could not escape (de Certeau 1984).

During the colonial era in Nicaragua, the indigenous and mestizo inhabitants began to perform a street play known as *El Güegüence*, which is still performed to this day. *El Güegüence* (see Figure 2.3) celebrates the creativity of indigenous and mestizo resistance to colonial domination.

Figure 2.3 *A figure from* **El Güegüence** *on display in the National Museum, Managua, Nicaragua*

Source: Kevin Glynn.

It shows how through clever language, mimicry and pretence of complicity, indigenous and mestizo actors were able to dupe and trick the Spanish colonizers, mock their authority and subvert their aims (Cuadra 1997). Through creative modes of resistance and cultural hybridization, indigenous cultures proved remarkably resilient and from the time of the conquest, native peoples have asserted their right to exist and in the process have often transformed and confounded the colonizers. While not denying the violence and brutality that underpinned colonialism, it is important to acknowledge the creative ways in which colonized subjects responded to the conditions in which they found themselves. Even when there was not outright rebellion or resistance, colonialism was never merely accepted intact.

Box 2.3

The Virgin of Guadalupe

A prominent example of cultural syncretism is the Virgin of Guadalupe, the patron saint of Mexico. Prior to colonialism, the indigenous peoples of the Valley of Mexico worshipped a maternal goddess known as Tonantzin. The Spaniards destroyed the temple built in her honour, replacing it with a Catholic shrine to the Virgin Mary. Worshippers continued to flock to the shrine, probably conflating the Catholic virgin with Tonantzin.

One morning in 1531, the Virgin Mary appeared in a vision to converted Indian Juan Diego on the Hill of Tepeyac, leaving an image of herself on his apron. She had brown skin like the native people and spoke in Nahuatl, and these factors along with her likeness to Tonantzin probably contributed decisively to the subsequent conversion of the indigenous peoples to Catholicism.

For Merrell (2004: 51) the Guadalupe–Tonantzin image constitutes one of the most evident sites of "the hybridization of Peninsular, African-American and Amerindian cultures" and, as he goes on to note, she did not only convince the indigenous peoples: the *criollos* (people of Spanish descent who were born in Latin America) also embraced her. The Catholic Church built a church in her honour at the site where she appeared, and Guadalupe has been central to Mexican culture and nationalism since that time.

Similar phenomena can be found in the Andes. For example, the carnival of Oruro in Bolivia is a hybrid religious festival in which indigenous customs have been blended with Christian ones. It involves a celebration of Pachamama, or Mother Earth, in which the locals worship a Catholic-style virgin, known as the Virgin of the Mineshaft, who, like Guadalupe, is said to have appeared in a vision, while Tío Supay (or the Uncle God of the Mountains) is blended with the Christian devil (Lecount 1999).

The independence movements

The colonial occupation of Latin America lasted 300 years and was characterized by ongoing indigenous and African slave resistance to colonial rule. Hundreds of rebellions were carried out by indigenous leaders and slaves. The most prominent of these include the rebellion of Atahualpa II in Bolivia in 1742 and that of Tupac Amaru II in Peru in 1781. Tupac Amaru rallied massive indigenous support when he laid siege to Cuzco, Peru, and issued a decree to end slavery, taxation and forced labour (Galeano 1971). In addition, the Spanish population was increasingly a native-born one, and these Spaniards became known as *criollos* or creoles, in distinction from the *peninsulares* – the settler population. Born in the New World, *criollos* had a different relationship to Spain and grew to resent the imposition of laws and taxes from afar. They felt such control was hindering their ability to trade and develop. At times, this resentment resulted in open insurrection and military conflict, such as the Revolt of the Comuneros in Paraguay in 1721–1735 and in New Granada in 1780–1781. Such demographic changes and the concomitant transformation of attitudes contributed to the gradual unravelling of colonialism and the intensification of the demand for independence.

There were however other important influences on the movements for independence. A successful slave revolt (1791–1804) in Haiti, a French colony, led to the abolition of slavery and the establishment of the Haitian Republic, the first independent black nation in the world. Enlightenment ideals, which disseminated the notion of progress and liberty, were influential in both the US and French Revolutions and also began to take root in Latin America. The desire for independence was further fuelled by the Napoleonic occupation of both Spain and Portugal, a move that caused the Portuguese monarchy to relocate temporarily to Brazil and undermined the allegiance felt by creoles in Spanish America to Spanish colonial rule. The political crisis and power vacuums created in Iberia by the Napoleonic invasion led to the creation of juntas in Latin America, made up of individuals who wished to maintain colonial rule as well as those seeking independence. The Napoleonic invasion of the Iberian Peninsula also provoked fears that France would gain control of Spain's colonies in the Americas.

The loss of political power was augmented by economic developments as Spain progressively lost its monopoly on trade with the Americas. It was struggling to provide adequate levels of supplies to the colonies and was coming under constant challenge from the British, French and Dutch, who all had colonizing and trading interests in the region. Britain, France and Holland cannot be excluded from the history of Latin America as they influenced the geopolitical configuration of the continent in a range of ways. Throughout the seventeenth century, these nations had established colonies in North America, Latin America and the Caribbean. In 1806 Britain invaded Buenos Aires, and in 1814 Guyana was taken from the Dutch by the British. In the absence of direct occupation and colonization, these nations also developed investment and trade linkages with the continent. By the 1800s, these challenges were joined by those from a newly independent United States.

The early 1800s saw the emergence of a number of independence leaders who waged bloody battles against Spanish troops. The most famous of these was Simón Bolívar (see Box 2.4 and Figure 2.4), others include

Figure 2.4 *A statue dedicated to Simón Bolívar in Bogotá*

Source: Pedro Felipe, Wikimedia Commons (Creative Commons Attribution-Share Alike 3.0 Unported).

Miguel Hidalgo, Francisco de Miranda, Manuel Belgrano, José de San Martín and Bernardo O'Higgins. Miguel Hidalgo, a creole who had plotted against the Spanish, is credited with beginning the move for independence in New Spain with his Grito de Dolores, a cry from the town of Dolores. Most of the independence fighters were male, but some women joined the struggle. In Colombia in 1817, the Spaniards publicly executed a 20-year-old woman, Policarpa Salvarrieta, who had rebelled against Spanish rule.

Most Latin American nations became independent between 1810 and 1825. Independence was not simply won and retained. In some cases, a republic would be founded that would then be reconquered by the royalists. The first countries to declare independence were Paraguay and Venezuela in 1811, although it was several years before independence was definitively secured. Other areas enjoyed a kind of *de facto* independence before independence was formally declared.

Box 2.4

Simón Bolívar, the father of South American independence

Simón Bolívar (1783–1830) was born in Caracas, the descendant of Spanish aristocrats. The Bolívar family amassed a fortune in the New World in silver and copper mining and sugar plantations. Bolívar, who received part of his education in Spain, dreamed of and fought for Latin American independence and unity. In his famous Letter to Jamaica, he posits the idea of America not as a set of distinct nation states but as a unitary whole.

After a number of battles against the royalists, Bolívar became the president of Gran Colombia, one of the first independent republics of Latin America, composed of the contemporary nations of Colombia, Venezuela, Ecuador and Panama. Along with José de San Martín, the liberator of Argentina, he also secured the independence of Peru. He helped to liberate Bolivia, and as a consequence the new nation was named after him.

In 1813, during a series of independence battles to liberate Venezuela, known as the Admirable Campaign, he was given a puppy, Nevado, as a gift by locals in the Venezuelan Andes. Nevado fought alongside Bolívar until he was killed in battle in 1821. Bolívar was ruthless in his drive for liberation, and urged for a "war to the death" that would involve the killing of *peninsulares* (those born in Spain).

Bolívar's struggle, while undoubtedly heroic and inspiring, suffered many setbacks. In many places royalist control was re-established after independence was declared, and Bolívar would have to retreat before launching a new set of attacks. Political instability and confusion were part of his presidency. Although he is regarded by many Latin Americans as a nationalist hero, he died a bitter man. He resigned his presidency and was planning to go into permanent exile in Europe, but died of tuberculosis before his departure at the age of 47.

It took several years before Spain recognized its former colonies' independence. France did not recognize the independence of Haiti for more than two decades, and did so in return for the imposition of a large debt burden. It also took several years for the nations that make up Latin America today to consolidate. Central America became independent as part of the independence of Mexico but later seceded to form the Federal Republic of Central America. Seventeen years later, the Federation collapsed and the contemporary Central American nations were formed. Panama was initially part of Colombia but with US help seceded in the early twentieth century. Belize failed to secure its independence with the other Central American nations. It was taken by Britain in 1836, named British Honduras, and remained a British colony until its independence in 1981. Puerto Rico and Cuba remained colonies of Spain until 1898.

Nation-building after independence

Independence did not lead to peace and stability for the region. Williamson (2009) describes independence as a zig-zag process made up of ongoing social inequalities and tensions, and confusion and disagreement over what mode of governance was most appropriate. While the struggles for independence led to the creation of self-governing modern nations, life did not improve substantially for the majority of Latin Americans. Indeed, the struggles were fought in large part by elite landowning groups who believed that they were superior to both blacks and Indians, and so these social groups continued to be subject to discrimination and marginalization. In other words, formal colonial rule by Spain and Portugal was brought to an end, but colonizing attitudes and structures of power remained in place.

Cuban revolutionary and independence fighter José Martí, described the period 1810–1825 as Latin America's first independence, but an independence that remained incomplete given the semicolonial dependence created by the extractive economy. He believed that Latin America required a second independence. In a much-celebrated essay entitled "Our America" (Martí 1977), originally published in 1891, Martí warned of impending US imperialism against which Latin America should unify and mobilize, and he called for the recognition and incorporation of local and indigenous knowledges.

Given these processes and the ethnic composition of the Latin American population, the post-independence period was a time in which the struggle for identity became paramount. By the early 1800s, most Latin

Americans were *mestizos* yet tended not to understand themselves in such terms. Indigenous and black populations continued to be viewed by whites as inferior peoples, whose capacity for civilization was in doubt. For the Eurocentric elites of the continent, who felt it was their role to construct modern, prosperous nations, there was a widespread concern that national progress would be hindered by the presence of so much indigenous and black blood, and the continent's racial mixture was frequently seen as the source of its social and economic failures. At this time, scientific racism, or a belief that the biological and hierarchical nature of race could be scientifically proved, was gaining ascendancy in Europe and beyond, and social Darwinism was the order of the day (see Chapter 6).

While the indigenous and African-descended populations continued to fight against oppression and internal colonialism in a range of overt and covert ways, national politics in many Latin American countries was characterized by battles between elite groups, who divided into liberals and conservatives. The colonial period left, in Black's (2011: 6) words, a legacy of "elitism, authoritarianism and militarism" which continued to shape the post-independence period, and Spanish ideals coexisted with more liberal ideas emerging from the European Enlightenment. While most nations became republics after independence, Mexico and Brazil experimented with monarchies first. Brazil gained its independence in 1822 but remained a monarchy for 70 years.

For the first half of the nineteenth century, political and economic instability reigned and conservatives and liberals argued over the best way to bring order to their unruly nations. The conservatives felt nostalgia for the Catholic monarchy and believed the power of the Church and the armed forces should be protected. Liberals believed in the sovereignty of the people, representative government and free markets (Williamson 2009). Despite their disagreements, as Williamson (2009) notes, both liberals and conservatives developed dictatorial and authoritarian tendencies, feeling that the chaotic situation in which they found themselves demanded strong leadership. Thus after independence, the phenomenon of *caudillismo*, or strong-man politics, emerged and became widespread throughout the continent. Williamson (2009: 237) defines a *caudillo* as "a charismatic leader who advanced his interests through a combination of military and political skills, and was able to build a network of clients by dispensing favours and patronage".

Latin America's most famous *caudillos* include Antonio López de Santa Anna of Mexico, Juan Manuel de Rosas of Argentina and Diego Portales of Chile. These were ruthless military and political leaders, with little tolerance of political dissent. They were, however, quite erratic ideologically. Santa Anna sided with royalists and republicans and with liberals and conservatives. He remains to this day a controversial historical figure and is seen as both a hero and a traitor. He clearly played an important role in Mexico's independence, but is also blamed for the drastic loss of Mexican territory to the United States in the 1848 war (see Fowler 2007).

Amid authoritarianism, clientelism and patriarchy, liberal ideals did make important inroads into continental political culture and the post-independence era also produced leaders such as Benito Juárez of Mexico and Domingo Sarmiento of Argentina who were in favour of democracy and social reform. And indigenous peoples and African-descended peoples continued to fight for their liberation from colonial and neocolonial structures of power.

By the second half of the nineteenth century, some countries had achieved a degree of political and economic stability and there were visible signs of economic development and modernization in some parts of the continent. International trade, which had fallen with the collapse of the Spanish empire, was expanding through increased exports and foreign investment in railways, mining and plantation agriculture. Countries such as Argentina and Uruguay were successfully attracting immigrants from Europe.

In this period, Latin America's role as an exporter of raw materials, created during the colonial era, intensified. Coffee, which had been brought to Latin America from Africa, began to boom owing to high demand in Europe. Brazil, Venezuela, Colombia, Costa Rica and Nicaragua became large exporters of coffee. Venezuela and Ecuador exported cacao, while Mexico expanded its silver production. Peru and Chile provided the guano and nitrates used in European fertilizers, while Argentina became a leading exporter of meat and wool. There were new export crops such as bananas, rubber, copra and chicle. Whole nations became reliant on one or two crops. Export crops tended to produce a boom-and-bust economy, creating great wealth at times for a minority. For the majority poor, converted into an exploited labour force, these commodities provided long hours of work and insecure income when they were booming and misery and hunger when their prices collapsed on international markets.

By 1920, Latin America was thoroughly integrated into the global economy. Despite some apparent stability after the turbulence of the independence struggles, the failure to address structural inequalities and in particular the question of unequal land tenure would continue to produce social tensions and would lay the pattern for the economic and political struggles of the twentieth century. Although most Latin American nations were independent by the 1820s, their independence was fragile, not least because of the economic rise of the United States. Indeed, the relationship between Latin America and an ascendant United States would have decisive effects for Latin American development in the coming decades (see Chapter 4).

Box 2.5

Nineteenth-century Mexico

While there are many similarities and interconnections between the independence struggles of Latin America, different parts of the continent faced different challenges. The route taken by Mexico has its own unique characteristics. The independence of Mexico was secured in 1821 via the Plan de Iguala, which contained three guarantees – independence, religion and union. These guarantees gave equal status to settlers born in Spain and creoles who were born in Mexico. But unlike other nations, Mexico would be a constitutional monarchy, ruled by Emperor Agustín de Iturbide. Both the monarchy and the concept of union collapsed very quickly. In 1823, Agustín was forced to abdicate and in 1828, Spaniards were expelled from the nation.

The newly created republic was constantly threatened by foreign invasion. Spain attempted to regain it, it was invaded by France, and in 1846 Mexico went to war with the United States and ended up being forced to cede a third of its territory to the United States in the Treaty of Guadalupe–Hidalgo, signed in 1848. For the first half of the century, liberals and conservatives blamed each other for this loss of territory and continued to fear US expansionism (Skidmore and Smith 2005). These elites were also in a constant battle over the role of the church and the question of sovereignty.

There were frequent changes of president, usually by military coup. For much of this period, Mexico was led by military general and *caudillo* Antonio López de Santa Anna, until he was overthrown in 1855. There was then an attempt to establish a secular state based on the rule of law and in which the church and the army were divested of their extraordinary privileges, a period known as *La Reforma*.

Mexico elected liberal Zapotec lawyer Benito Juárez as its first civilian president. A further occupation by France in 1864 meant another brief experiment with constitutional monarchy as Archduke Maximilian von Hapsburg came to power, but he was killed by firing squad one year later. The liberals were returned to power but *La Reforma* soon gave way to *El Porfiriato* when Porfirio Díaz seized power. During his long, highly centralized and almost uninterrupted presidency (1876–1910), the Mexican economy grew rapidly. This growth was achieved through the consolidation of mining and the expansion of export agriculture. In the process, indigenous peoples and poor rural *mestizos* found themselves further marginalized.

Mexico was modernizing rapidly but the majority of Mexicans had to endure hunger, landlessness, infant mortality and misery, and these were the roots of the Mexican revolution which began in 1910. Revolutionary leaders such as Emiliano Zapata and Pancho Villa were able to mobilize landless peasants and other dissident sectors of society and brought about radical political change, including land reform and the protection of workers' rights, enshrined in the Constitution of 1917.

Conclusion

As we shall see in the next two chapters, the legacies of colonialism and the mode in which independence was established continued to shape Latin America's economic and political development well into the twentieth century. The four centuries between the Iberian conquest and the start of the twentieth century cannot be easily encapsulated in a single chapter. They were tumultuous times, in which some indigenous cultures were dramatically destroyed but others incredibly managed to survive. Export agriculture became entrenched, bringing both wealth and poverty depending on one's social position. Extreme forms of masculinity were often privileged, and Latin America's historical record has provided a number of somewhat controversial heroes.

By the start of the twentieth century Latin America was economically, politically and culturally diverse; it would become more so in subsequent decades. The struggle between wealth and poverty, between privilege and misery, would continue. Inequalities of wealth, gender and race would give way to some of the most intense struggles over development.

Summary

- Pre-Columbian Latin America was home to many different cultures and cultural practices.

- In the fifteenth and sixteenth centuries, what is now Latin America was colonized by Spain and Portugal.

- Colonialism was violent and brutal and inflicted great harm on the indigenous populations.

- Colonial legacies can be observed in languages, religion, agriculture and land tenure, architecture, urban form and political culture.

- Colonialism converted Latin America into a supplier of raw materials which fuelled the development of Europe and later the United States.

- Colonialism was resisted in many ways by indigenous, African, mestizo and Spanish settler populations and led to independence struggles through the continent.

- Most Latin American countries became independent in the nineteenth century.

Discussion questions

1. What were Spain's principal motivations when it conquered what is now Latin America?

2. How did indigenous peoples adapt to and resist colonial rule?

3. What kind of foreign policy relationships were established between Latin America and the United States in the nineteenth century?

4. What are the main political and economic legacies of the conquest? What political cultures and models of economic development did the conquest facilitate?

5. Why did some intellectuals feel that Latin America's independence from Spain and Portugal was deficient or incomplete?

6. How did the post-independence history of Mexico differ from that of other Latin American nations?

Further reading

Bethell, L. (ed.) (2009) *The Cambridge History of Latin America*. Cambridge: Cambridge University Press. A 12-volume set that comprehensively surveys the history of Latin America from 1500 to the present.

Burkholder, M. A. and Johnson, L. L. (2009) *Colonial Latin America*, 7th edn. Oxford: Oxford University Press. A good history of the colonial period.

Chasteen, C. (2011) *Born in Blood and Fire: A Concise History of Latin America*, 3rd edn. New York: W. W. Norton. A valuable and engaging introduction to Latin American history.

Mills, K., Taylor, W. B. and Lauderdale Graham, S. (2002) *Colonial Latin America: A Documentary History*. Lanham, MD: Rowman & Littlefield. A sourcebook of primary documents and visual material pertaining to colonial Latin America.

Restall, M. (2004) *Seven Myths of the Spanish Conquest*. Oxford: Oxford University Press. This book tackles some of the common misunderstandings and misrepresentations of the Spanish conquest of Latin America.

Skidmore, T., Smith, P. H. and Green, J. (2009) *Modern Latin America*, 7th edn. Oxford: Oxford University Press. Using a comparative analysis and case study approach, this is a valuable introduction to historical and political trends in Latin America.

Williamson, E. (2009) *The Penguin History of Latin America*. London: Penguin. A concise yet comprehensive introduction to the history of Latin America.

Films and documentaries

Central America: History and Heritage (1985), directed by Wayne Mitchell (USA). A documentary that explores the history and colonization of Central America.

Simón Bolívar, Ese Soy Yo (1994), directed by Edmundo Aray and Raiza Andrade (Venezuela). An animated portrayal of the South American liberator.

El Pueblo Mexicano que Camina (1997), directed by Juan Francisco Urrusti (Mexico). A Mexican documentary that explores the cult of Guadalupe.

Aztecs: Inside the Hidden Empire (1999), directed by Mark Gray and Liz Hamill (UK). A documentary that explores Aztec civilization, beliefs, culture and violence. www.youtube.com/watch?v=G4JabdIKx8s

3 Economic development
The rise (and fall?) of neoliberalism

Learning outcomes

By the end of this chapter, the reader should:

- **Understand different economic approaches to development, including modernization theory, dependency theory, import substitution industrialization and structural adjustment**
- **Understand the factors that contributed to the debt crisis of the 1980s**
- **Be able to evaluate the impacts of structural adjustment policies on different sectors of the population and the economy**

States and markets

During the nineteenth century, competing narratives about how the newly independent nations of Latin America should be organized economically, politically and culturally were in circulation. The classical economic theories espoused by Scottish economist Adam Smith in *The Wealth of Nations* along with David Ricardo's theory of comparative advantage had become extremely influential in Europe and beyond, and had brought to an end the mercantilist model that had dominated global trade. Classical economics promotes free markets and specialization through an international division of labour. This body of economic thought was appealing to Latin American liberal elites and the new entrepreneurial class that emerged after independence. The question of free trade was particularly central to Latin American economic development, given that independence movements had fought in part to free up trade from the monopolies held by Spain and Portugal. Although debates about the role of the state and the value of statism and protectionism in generating prosperity continued to circulate, by the start of the twentieth century, free trade thinking clearly had the upper hand. It was well supported by local entrepreneurs and foreign investors and underpinned by a credible body of economic thought.

By 1900, the commodity trade with both the United States and Europe had intensified and Latin America was well integrated into the global economy. But the problem with this form of growth was that it relegated Latin America to a peripheral or dependent status, in which its economic role was that of producing raw materials to fuel the economic growth and industrialization of Europe and the United States. Consequently, Latin America was forced to use the foreign exchange that it earned to import industrial and manufactured products. Nonetheless, growing global demand for Latin American commodities meant that for both local and foreign economic elites there was great wealth to be made in the export economy, particularly in beef, wheat, sugar, coffee, oil, copper and bananas (Skidmore and Smith 2005).

The liberal free trade model ran into difficulties following the crash of the Wall Street stock exchange in 1929. The crash ended a speculative bubble and led to a global economic depression and mass unemployment in much of the world, which lasted until the end of the Second World War. While the recession was a global one, it had very particular effects and consequences in Latin America. Commodity prices slumped and Latin America's export markets all but disappeared overnight, and with them the foreign currency Latin America used to purchase imported manufactured goods. Latin American economies were forced to adapt very quickly to a new reality. Poverty, misery and social unrest increased. At this point, free trade and the export-oriented model of development were practically and ideologically exhausted. There was a clear sense that the model had failed and some economic rethinking was urgently required. In particular, it became apparent that Latin American governments needed to begin to protect their economies and produce for themselves many of the manufactured goods that were being imported.

In 1944, towards the end of the Second World War, the institutions of global governance, the International Monetary Fund and the World Bank, were created at Bretton Woods to regulate the global economy and rebuild the war-torn countries of Europe (see Box 3.1). At that moment, development *qua* modernization was being promoted by the United States and other nations, and the countries of Latin America began to be viewed as lacking and requiring expertise from the developed countries. This was also the period in which Keynesianism came to the fore. British economist John Maynard Keynes, who played a key role in the Bretton Woods conference, was in favour of private enterprise, but believed that states should actively use fiscal and monetary policy to stabilize and stimulate the economy. Governments should control the money supply and therefore inflation through the setting of interest rates. In times of depression or stagnation, they should spend public money to activate the economy.

Keynesian thinking became influential in Latin America as it did throughout the world, but was extended through geographically specific economic thought. Argentine development economist and president of the UN Economic Commission for Latin America (ECLA), Raúl Prebisch (1950), theorized that the free market model which had dominated Latin America in the nineteenth century could not lead to economic independence for the region. He had observed that during the depression the prices of raw materials had fallen more dramatically than the prices of manufactured goods, a state of affairs economists refer to as unequal terms of trade. He believed Latin American nations needed to undergo their own processes of industrialization and the state needed to play a central and decisive role in directing the economy away from the production of commodities such as minerals and agricultural goods. The new local industries would need to be protected from international competition to allow them to flourish.

Box 3.1

The Bretton Woods institutions

The International Monetary Fund (IMF) and the World Bank were created near the end of the Second World War in Bretton Woods, New Hampshire, at a conference held by Allied governments with a view to rebuilding and regulating the global economy in the wake of conflict and destruction. In particular, the Allies wanted to avoid a repeat of the conditions that had led to the Depression of the 1930s and the subsequent Second World War. In the interwar years, through a policy referred to as "beggar-thy-neighbour", countries devalued their currencies and imposed tariffs in ways that harmed their neighbours, so delegates were aware of the need for greater economic collaboration between nations. The member countries agreed to a system of fixed or "pegged" exchange rates using the US dollar (which at the time was still backed by gold) as a reserve currency. The US dollar thus began to play the role in the international financial system that was previously played by gold.

This system remained in place until 1971, when a combination of factors including the US economic strategy, the Vietnam War and the oil crisis made it untenable. There was a move to floating exchange rates but the US dollar (no longer backed by gold) retained its hegemonic position.

When created, the IMF and the World Bank had quite different roles and mandates. The IMF was charged with overseeing global monetary policy and would lend to member countries experiencing balance of payments difficulties. The World Bank was a development bank that provided loans to member countries for development or reconstruction projects. These institutions were created not on a "one nation, one vote" system (as in the United Nations) but on a "one dollar, one vote" system. The more economically stable and wealthy US, by contributing much larger funds, came to dominate the IMF, possessing the ability to veto any policy measure that it did not support.

Over time the role and nature of these institutions have changed dramatically and the distinction between them has become less clear. In the post-war period, the World Bank made development loans to Latin American countries, in a manner that supported US geopolitical interests in the region (Woods 2006), and the IMF had balance of payments programmes in a number of countries in the region. But it was not until after the debt crisis in 1982 that these two institutions got involved in the management of Latin American economies in an all-encompassing and dramatic way. Their involvement in promoting neoliberal structural adjustment policies across the continent would set the pace of Latin American development for decades to come.

The IMF and the World Bank have been the targets of high-profile activism and critical scholarship, and are widely seen to have had devastating effects on the lives and livelihoods of ordinary Latin Americans.

Consequently, a new form of development came into being, known as import substitution industrialization or ISI. It was a model in which states actively intervened to promote economic development rather than leaving it to the invisible hand of the market. The governments used import tariffs to protect local industry from foreign competition and embarked on the nationalization of key industries such as iron and steel production. Electricity, mining, telecommunications, transportation and finance were all areas in which there was substantial state ownership and control. They also implemented price controls, maintained an overvalued exchange rate, and subsidized basic goods such as foodstuffs and utilities such as electricity to keep them affordable for the majority. National development banks with preferential interest rates for national investors were created to provide public finance for industrial investment. In Latin America, this mode of economic thinking is sometimes referred to as structuralism and forms the basis of dependency theory (discussed below). During this period, Latin America did however continue to export commodities, particularly as global demand for its commodities from industrialized countries increased once again during the Second World War (Green 2006).

ISI had some clear advantages and disadvantages for Latin American nations. As Green (2006) states, some of its results were impressive, particularly in the larger nations such as Mexico and Brazil. Gross Domestic Product (GDP) tripled, and Latin American economies grew more rapidly than the first world economies of Europe. By the early

1960s Brazil had become self-sufficient in the production of manufactured goods such as televisions, refrigerators and electric stoves (Green 2006). High import tariffs also encouraged multinational companies to initiate production within Latin America, so in this period Brazil and Mexico became major producers of GM, Ford, Volkswagen and Fiat cars (Franko 2007). The standard of living also improved. Infant mortality fell and life expectancy increased. There was some public investment in literacy programmes, worker education and housing. In some countries, such as Argentina and Chile, industrialization combined with populism facilitated the creation of organized labour.

Despite some evidence of social and economic progress, ISI did not prove to be sustainable over the long term. Protectionism allowed domestic industry to grow but also encouraged inefficiency and corruption. Many ISI products were expensive and of poor quality, and the lack of external competition meant there was little incentive to innovate. The ISI model was good for the cities, but devastating for the rural areas. Because government kept food prices artificially low and concentrated public investment in the cities, small farmers found making a living increasingly difficult. From the 1950s onwards, vast numbers of rural people migrated to Latin American cities, in search of employment and hoping for a better standard of living. But formal sector employment was in short supply and so many people were condemned to insecure and precarious work in the informal sector. Local governments were unable to keep up with the demands for infrastructure, so many migrants found themselves living in makeshift shelters in shanty towns on the edges of cities (see Figure 3.1), without services such as water, electricity, schools and public transport, and without the income to purchase consumer goods (and therefore contribute to economic expansion). So while ISI created some wealth, it was not evenly distributed and both misery and inequality continued to characterize the Latin American economy.

Although industrial growth did take place in some parts of the continent, Latin America continued to be a key exporter of primary products. In addition, ISI could not be fully implemented because Latin America still needed to import capital goods from Europe. Steel production, for example, was dependent on the importation of furnaces (Franko 2007). While Brazil and Mexico had developed a viable car industry, these companies remained foreign-owned and would therefore expatriate profits.

The forms of economic management under ISI generated other economic problems, including high inflation. Governments would print money to

Figure 3.1 *A shanty town or* villa miseria *on the outskirts of Buenos Aires*

Source: Aleposta, Wikimedia Commons (Creative Commons Attribution-Share Alike 3.0 Unported, 2.5 Generic, 2.0 Generic and 1.0 Generic).

finance expenditure, which in the absence of real income growth created inflationary pressures. Some governments in the region began to implement heterodox stabilization measures in an effort to curb inflation. In 1980s Nicaragua, the government printed money to finance the military struggle against the US-backed Contra forces, and consequently inflation rose to 10,000 per cent by 1988 (Cardoso 1989). Capital flight was also a serious problem, not just through the multinational repatriation of profits but also because wealthy Latin Americans preferred to keep their money in US bank accounts and currency overvaluation at home made it profitable for them to do so.

Retheorizing economic development

The implementation of ISI in Latin America coincided with the rise of modernization thinking, discussed in Chapter 1, and the establishment of development studies as a specific field of enquiry. While the structuralists and the modernization theorists differed in their understanding of the causes of underdevelopment, they were not at odds with one another at this time. In the immediate postwar period, the United States administration was not concerned about the existence of ISI in Latin America, a position that would later change. The structuralists also shared modernizing visions. They were motivated by notions of development as progress and willing to take on large loans to finance large-scale development projects such as dams and steel plants.

By the 1960s, both the ISI model and faith in modernization were starting to unravel. Latin America, despite some incipient industrialization, continued to be peripheral to the world economy. The promised benefits of development were failing to arrive, especially for the poor majority in both cities and rural areas, and it was clear that Latin America could not follow the same economic trajectory taken by Europe in earlier periods. Structuralist critiques would evolve into what has come to be known as dependency theory.

Dependency theorists such as German André Gunder Frank, Chilean Osvaldo Sunkel, Brazilians Theotonio dos Santos and Fernando Henrique Cardoso (who later became the president of Brazil) along with Prebisch articulated a well-developed critique of modernization theory. They asserted that modernization and the concomitant attempt to integrate third world countries into the global economy rather than leading to development was *causing* underdevelopment and dependency. They blamed the economic legacies created by the colonial system of commodity trade which condemned Latin American countries to ever deteriorating terms of trade, in which the increase in prices of manufactured goods constantly outpaced the increase in prices of commodities, forcing the production of more and more units of raw materials to maintain the same levels of imports of manufactured goods. In other words, the development of the core was causing the underdevelopment of the periphery and the poor of Latin America were contributing to prosperity in Europe and the United States, while they reproduced their own disadvantage. This body of theory was influenced by Marxist thought, and dependency theorists to varying degrees advocated breaking relations with the advanced capitalist nations and promoting socialist revolution and self-sufficiency for Latin American nations.

While dependency theorists in Latin America and elsewhere were formulating Marxist-inspired analyses of the global economy, free market capitalism was starting to experience an intellectual and political revival elsewhere in the world. The Chicago School of Economics under the leadership of Milton Friedman and Arnold Harberger was staunchly advocating a return to free market economics, arguing that state-led economic development had been a failure. These intellectuals had a strong influence in Latin America, especially in Chile in the late 1970s. Guided by the Chicago Boys, a group of Chilean economists who had studied at the Chicago School, the Pinochet dictatorship began to implement radical free market reforms.

The debt crisis

With the exception of Chile, the ISI model lasted in various forms until the early 1980s, amid growing dissatisfaction with its results and those of development more generally. The global economy changed dramatically in the 1970s in ways that would have long-lasting effects in Latin America and would lead to yet another dramatic economic transformation and the total dismantling of the ISI model.

Until the 1970s, the industrialized countries of Europe and the United States benefited from Keynesian economic policies. In addition to economic growth, Keynesianism had facilitated massive public investment in education and health care that had improved the standard of living in the first world. In 1973 and again in 1979, the oil-producing countries (OPEC) dramatically increased the price of oil – a move that produced stagflation (simultaneous stagnation and inflation) in oil-importing nations such as the US and the UK. Keynesian demand management had always proceeded to correct either a depressed or an inflating economy, but in the 1970s these phenomena appeared simultaneously – a situation for which there were no Keynesian tools.

While western nations were stagflating, the OPEC countries were making massive profits. These profits, known as "petrodollars", were often deposited in commercial banks in both the US and the UK. Given the economic situation in their countries, the bankers were desperate to recycle the petrodollars and they aggressively offered loans to developing countries, especially in Latin America. In what became known as the dance of the millions, a Citicorp or Midland Bank banker would turn up in Argentina or Brazil and provide an instant multimillion dollar loan to the government. At this time, many countries in Latin America were ruled by military dictatorships that spent the funds in unproductive ways – on steel plants that never produced any steel, on dams that destroyed the surrounding environment or on weapons that they used against their own people. Some of the funds, however, were used to finance development, creating new industries and sources of employment, building roads, extending the provision of electricity and developing telecommunications. By 1982, Latin America had borrowed $300 billion from first world banks (Franko 2007). Despite the scale of the borrowing, debts appeared sustainable to both lenders and borrowers, as global interest rates were low – indeed they were negative in real terms – and commodity prices were high. Creditors were confident that debts could be repaid with export revenue.

This situation began to change after the oil crises and the rise to power of neoliberal global leaders such as Margaret Thatcher in the UK and Ronald Reagan in the US. Thatcher and Reagan implemented restrictive monetarist policies in an attempt to deal with the stagflation afflicting their economies, hiking up global interest rates. Also, the oil crisis had resulted in a fall in the price of commodities. Suddenly, the gap between what Latin American countries were able to earn from their exports and what they were required to pay to service their debts widened. This was the beginning of what came to known as Latin America's lost decade.

The debt crisis began on Friday, 15 August 1982, when Mexican finance minister Jesús Silva Herzog discovered that that the Mexican government did not have sufficient funds to make the debt service payments due on the following Monday. The government had run out of foreign exchange reserves. The threat of default by Mexico sent the first world bankers into panic. Many had lent more than 100 per cent of their shareholder capital to governments in Latin America and elsewhere. They knew that if the default was to be repeated across the developing world, it would lead to the collapse of the global financial system. In order to save the banks and prevent the formation of a debtors' cartel, the IMF was appointed as the overall crisis manager and the banks were forced by the IMF to contribute to an emergency rescue package for Mexico. Consequently, Mexico was bailed out – it was offered new loans to keep its debt service payments flowing. The same negotiations were repeated bilaterally with many countries across Latin America. At all costs, the banks had to prevent debts from going bad. Paradoxically, the solution to unpayable debts was to extend the overall debt burden carried by debtor countries. By the end of the 1990s, Latin America's debt burden was almost three times as large as it had been in 1982 and the crisis had devastating consequences for ordinary Latin Americans.

Although the Mexican default had sent the first world bankers into a tailspin, for many economists the debt crisis was a welcome opportunity and provided the catalyst needed for a move back to the free trade model that had been so discredited in the 1930s. The IMF, with support from the commercial banks and the US Treasury, took advantage of the situation and, underpinned by a growing body of economic theory coming out of prestigious schools, used the debt crisis to demand a dramatic overhaul of the economies of Latin America in line with neoliberal thinking. Countries were offered fresh loans with which they could meet their debt service requirements, but in return were expected to comply with a stringent set of economic conditions, known as IMF conditionality – conditions that would bring the ISI model to an end.

IMF conditionality varied from country to country but generally contained a mix of the following policy ingredients: a cut in public spending, promotion of exports, the elimination of government subsidies, currency devaluation, privatization of state-owned enterprises, and the liberalization of foreign trade and investment. It amounted to a substantial reversal of economic policy from one based on inward-focused development and protectionism to one based on exports and liberalization. This approach became known as structural adjustment and, over the course of the 1980s and early 1990s, most Latin American countries fell subject to IMF conditionality. The support for such policies from the US government and powerful institutions based in Washington, DC meant that the policy package became known as the Washington Consensus.

In 1995, the IMF and the World Bank were joined by a third powerful multilateral organization, the World Trade Organization (WTO), which added further momentum to their economic prescriptions. While the WTO maintains it is a neutral site in which member states can resolve trade disputes, its ideological position nonetheless is "passionately against protectionism and just as profoundly for trade liberalization" (Peet 2003: 160).

While the IMF recipe has a number of common features, it is important to recognize that the timing, rate and degree to which this set of policy prescriptions was implemented varied quite markedly from one country to another. Neoliberalism is not a monolith and instead should be understood as "a complex assemblage of ideological commitments, discursive representations, and institutional practices, all propagated by highly complex class alliances and organized at multiple geographical scales" (McCarthy and Prudham 2004: 276).

It is also important not to paint a picture of an all-powerful IMF imposing its will on weak developing nations that had no leverage or bargaining power. In some cases, such as in Mexico in the early years after default, as Woods (2006) has noted, structuralist and nationalist positions within the government were resilient, cabinets were heterodox and the IMF had to bargain hard to get an agreement in place. But after the election of Carlos Salinas de Gortari in 1988, Mexican officials became closely aligned with the policy prescriptions of both the IMF and the World Bank, and a high level of cooperation developed. Despite the implementation of an orthodox neoliberal model in the 1970s, the Chilean economy also went into crisis in 1982. As a result, neoliberalism became much more pragmatic and heterodox between 1982 and 1989 (Ffrench-Davis 2010). In other countries, government officials and the general public were frequently quite receptive to such policies from the start, given high inflation, persistent poverty and the general state of national economies. An opinion poll conducted in Peru in 1988 showed that 75 per cent of those polled supported an agreement with the IMF (Cardoso 1989).

Neoliberal supporters often tend to present their policies in a "there is no alternative" (TINA) mode, acknowledging that the medicine might be painful but it is the only possible option. In the midst of debt default, fiscal deficits and spiralling inflation, it did seem at the time like there was no alternative. Neoliberalization has been a complex, uneven and at times contradictory process in Latin America and it has had its own local proponents. In Peru, the work of economist Hernando de Soto and the Institute for Liberty and Democracy have endorsed many of the goals of neoliberalism but have also supported small farmers and informal sector workers to gain land titles or formal business registration (see de Soto 1989).

Panizza (2009: 10–11) says it is useful to think of neoliberalism as "a relational construct, whose contested meaning is defined and redefined by the political struggles between its defenders and detractors, and to which specific policies are contingently articulated according to the history and contexts in which they are fought". It is useful therefore to gain insight into the general principles on which IMF conditionality is based but also to study its implementation in specific contexts and the ways in which neoliberal economic policy is negotiated by both policymakers and ordinary citizens.

Impacts of neoliberal structural adjustment

The implementation of neoliberal structural adjustment policies had devastating social and environmental outcomes. The IMF measures resulted in a dramatic fall in real wages and an increase in poverty and unemployment as trade liberalization wiped out local businesses and local markets were flooded with cheaper imported goods, including staples such as corn and rice that were also grown locally. People were also harmed by cuts to health care and education, and the elimination of subsidies on food, fuel and transport dramatically increased the cost of foodstuffs such as tortillas, bread and milk and the price of bus fares to work or school. The United States continued to subsidize its own agricultural production heavily, even as it insisted that Latin America end its subsidies, which made it impossible for Latin America to compete with the US. The new emphasis on export crops meant that small farmers found themselves deprived of the bank credit they needed to buy agricultural inputs, and many lost their land as a result of the pressure from large agribusiness. It also undermined local food security, as many small farmers who grew crops for domestic consumption were no longer able to plant and some took jobs on export plantations or migrated to the cities where many people, including children, work in the informal sector. A report by British aid organization Oxfam described the effects of structural adjustment during the lost decade as follows:

> In Latin America, per capita incomes dropped by 10 per cent and investment fell from 23 per cent to 16 per cent of national income during the 1980s. Import activity dropped sharply, as governments transferred a massive stream of wealth – totalling over $200 billion, or some 6 per cent of GDP, during the decade – out of the region. Inevitably, deflation on this scale increased social misery, with the costs of adjustment passed disproportionately to the poor. The World Bank itself estimates that nearly one third of the region's population was living in poverty by 1990, up from 27 per cent a decade earlier; and an estimated 10 million children are suffering from malnutrition.
>
> (Oxfam Policy Department 1995: 8).

Until well into the 1990s, many countries were spending more money on debt service than they were on education (see Figure 3.2) and healthcare (see Box 3.2). Taking Latin America and the Caribbean as a whole, Potter (2000: 66–7) shows how the region paid US$1.65 trillion in debt service between 1980 and 2000, and still owed US$750 billion in 2000, up from US$191 billion in 1981 – a situation that she describes as "the more we pay the more we owe".

As we shall see in Chapter 5, the environmental consequences of adjustment have also been catastrophic. Many key Latin American assets passed into both foreign and private hands. The privatization of national industries enabled US and European multinationals to gain substantial financial interests in Latin America's airlines, telecommunications, oil and electricity. The privatization of water and electricity has in many cases failed to deliver efficiencies and has led to higher prices for

Figure 3.2 *NGO support for education, Quetzaltenango, Guatemala. A Spanish NGO, Intervida, helps to compensate for the education deficit created by the external debt*

Source: Marney Brosnan.

consumers (see Cupples 2011). Large private investors were able to make windfall profits while ordinary Latin Americans were struggling to put food on the table and began to pawn their belongings.

Latin Americans, young and old, male and female, had to develop creative strategies in order to survive the economic downturn, many of which were not developmental. Children were pulled out of school so they could clean shoes or windscreens at traffic lights (see Figure 3.3). Older children became the caregivers of younger siblings, while their mothers walked the streets selling lottery tickets or home-baked goods. Schoolteachers supplemented meagre incomes by taking in washing and ironing from more affluent families. In some areas, outmigration increased, with mixed outcomes for development (see Box 3.3). Studies began to reveal how low-income women were particularly negatively affected by adjustment, as the cuts to welfare and food subsidies made caregiving so much harder (Elson 1991; Benería and Feldman 1992; Moser 1993; Sparr 1994; Zack-Williams 2000; see Chapter 6).

One factor did however provide an economic cushion in times of austerity, and that was the trade in illegal drugs. While the trade creates

Figure 3.3 *Many children are forced to work in Nicaragua's informal economy. Shoe-shining is a common occupation*

Source: Marney Brosnan.

addiction and violence, its profits also find their way into the legal economy and development-related initiatives. During Bolivia's structural adjustment, coca production absorbed many workers such as tin miners who had become newly unemployed. While coca is a legal and traditional crop in Bolivia, much of the coca harvest is converted into its illegal and highly lucrative derivative, cocaine (see Leóns and Sanabria 1997).

Box 3.2

Structural adjustment and health

When structural adjustment policies were first imposed on Latin American countries, they dramatically reduced social spending on health, education and housing. Such policies left millions of Latin Americans without access to basic health care. Neoliberal health reform is quite difficult to achieve in many ways, because of the inherent unprofitability of disease and because many oppose on moral grounds the idea that it is acceptable for one person to profit from another's ill health.

Health care has not however been immune to market logic and commodification, and it is seen by proponents of neoliberal thinking as a potential source of capital accumulation. Under neoliberalism, health care is partially transformed from a social right to a market commodity and user fees were introduced in some places (Muntaner, Guerra Salazar, Rueda and Armada 2006). In some countries it has led to a dual and stratified system of private and public health care, a system promoted by agencies such as the World Bank and underpinned by loans for health sector reform.

Laurell (2000) describes how Chile developed a private system for a minority able to pay and with few health risks alongside a far more poorly resourced public system for the majority poor and the sick. She notes how such a system undermines solidarity between the two groups, reverses the trend towards universal health care provision and redistributes scarce resources towards private insurance companies and elite groups.

In Mexico and Brazil, neoliberal reforms put the public system under extreme pressure so that it could not meet the needs of low-income patients, while local and international insurance companies enjoyed substantial profits (Muntaner *et al.* 2006). Like other aspects of neoliberal policy, neoliberal health reforms are coming under challenge by "pink tide" governments. The idea that health care is a human right that should be publicly funded out of taxation has been enshrined in Venezuela's Bolivarian Constitution and the Venezuelan and Cuban governments have created a collaborative health care model to provide primary, secondary and tertiary levels of health care to low-income populations (Muntaner *et al.* 2006).

Box 3.3

Migratory flows and remittances

One of the key outcomes of economic hardship is migration. Since the Second World War, Latin America has produced large numbers of migrants who leave their region of origin to seek employment or educational opportunities elsewhere. Many migratory flows have been fuelled by intensifying misery in rural areas and have led to the creation of shanty towns or informal settlements on the edges of all major Latin American cities, as small subsistence farmers and their families were displaced by large landowners and export crops, environmental degradation and lack of available credit.

Many rural and urban Latin Americans also migrate to neighbouring countries, both illegally and legally. There are for example many Mexicans in the United States, many Nicaraguans in Costa Rica and many Bolivians in Argentina. Many make substantial sacrifices to migrate, sometimes hiring *coyotes* to get them across the desert that spans the US–Mexican border. Some never make it and perish *en route*; others are picked up by the US border patrol and are immediately returned to their countries of origin. When they do succeed, they often take on dirty, poorly paid and precarious jobs that existing residents do not want to do.

Migration has a number of effects, both positive and negative, on the migrants themselves and on the places that they leave behind and their adopted countries. It is usually the young and most productive members of society that migrate, leaving children and the elderly behind, so areas of high outmigration can lose their economic vibrancy. On the other hand, migrants often send remittances to their families at home which can enable them to feed, clothe and educate children, remain on their land and invest in development initiatives. Remittances from Latin American migrants are substantial and far exceed both foreign aid and foreign direct investment. According to the Inter-American Development Bank, remittances to Latin America and the Caribbean exceeded US$60 billion in 2011. In some countries, including Haiti, Guyana, Honduras, El Salvador and Nicaragua, remittances account for more than 15 per cent of GDP (Maldonado, Bajuk and Hayem 2012). To maximize the benefits of remittances for Mexico, the Mexican government created the 3 for 1 programme. Mexicans living in the United States are encouraged to create hometown associations (HTAs) and send money to their communities of origin. There are around 3000 Mexican HTAs in the United States (Vargas-Lundius, Lanly, Villareal and Osorio 2008). Every dollar sent is matched by $3 from the Mexican government and is spent on community development projects such as schools, potable water and electrification.

Migrants' experiences of migration are diverse. They frequently gain new skills and knowledges and are freed of some of the anxieties that they endured back home. At the same time, they are often subject to discrimination or racism. Being undocumented, as many Latin Americans in the United States are, can compound a sense of marginalization. The economic recession in the United States has been accompanied by a wave of anti-immigrant sentiment, forcing Latin Americans to mobilize in defence of their human rights.

The criticisms of structural adjustment from ordinary people, aid agencies, activists and scholars both within and beyond Latin America began to mount. Organizations such as Jubilee 2000 called for the cancellation of third world debt; others such as 50 Years is Enough called for the abolition of the IMF and the World Bank. Concerned in particular by the impact of adjustment on children, UNICEF began to call for "adjustment with a human face" (Cornia, Jolly and Stewart 1987). The scholarly and activist literature that outlines the harm done to ordinary Latin Americans by the IMF and the World Bank is enormous! Indeed, some of the former proponents of the Washington Consensus, including IMF and World Bank economists, began to distance themselves from their earlier beliefs and have publicly acknowledged that their policies in Latin America and elsewhere in the third world have failed. In 1988, an IMF economist, Davison L. Budhoo, who quit the Fund and published his 150-page resignation letter (Budhoo 1990), condemned the harm done by IMF policies, including the way in which statistics were manipulated in order to push damaging structural adjustment policies on third world countries (Klein 2007). Joseph Stiglitz was a chief economist at the World Bank and was dismissed in 2000 for his open criticism of structural adjustment policies (Stiglitz 2002). Even Jeffrey Sachs, who helped the Bolivian government to create its shock therapy in 1985 (see Box 3.4) has shifted his position considerably to become a staunch advocate of foreign aid, although as Peet (2007) argues, his criticisms of the neoliberal policy regime do not go anywhere far enough and continue to be ethically problematic and environmentally deterministic.

Box 3.4

Adjusting Bolivia

In 1985, Bolivia was considered to be a basket case by the international community. The government, unable to secure sufficient revenue through taxation, simply printed money to pay for the things it needed. As a result the country's inhabitants were confronting the daily reality of hyperinflation which reached a staggering 23,000 per cent – a rate which meant that prices increased every hour. The heterodox stabilization measures implemented after the debt crisis were showing no signs of working.

After the 1985 election, young Harvard economics professor Jeffrey Sachs flew to Bolivia to draft a drastic economic plan with the Minister of Planning, Gonzalo Sánchez de Lozada, which brought in free market reforms, a plan guaranteed to gain IMF approval. Price controls were eliminated, public spending was slashed and trade was liberalized. To get the harsh and controversial measures through with limited opposition, the plan was negotiated in secret and then approved by presidential decree 21060, and the president declared a state of siege so that workers could be fired and union leaders silenced.

Hyperinflation was resolved and eventually the Bolivian economy returned to growth, and these positive economic indicators meant that Bolivia could be held up by the IMF and World Bank as a reform success. Yet the price paid for such "success" by Bolivians was extremely high (Green 2006). The elimination of price controls put foodstuffs out of the reach of many ordinary Bolivians, malnutrition increased and thousands of workers were laid off. The downsizing of the state-owned tin mining company brought economic misery to thousands of miners and their families. Poverty was never resolved, and the continued failures of adjustment led to a decisive rejection of the neoliberal model in Bolivia. In May 2010, Bolivia's first indigenous president, Evo Morales, announced the creation of a new decree that would bring about the complete elimination of 21060.

Despite growing criticism, throughout the 1980s and into the 1990s, the IMF and the World Bank continued to assert that their policies were the only viable option for development and they repeatedly pointed to what they saw as evidence of reform successes. But by the mid-1980s it was clear that structural adjustment was failing to bring the promised benefits. Latin American countries still had growing and unpayable debt burdens and ordinary people were suffering. While inflation was brought under control in some parts of the continent, in others it continued to remain excessively high despite the economic measures. In Argentina, the government embarked on intense adjustment in the late 1980s but still could not bring inflation under control (Panizza 2009). Over the past three decades there have been a number of attempts to resolve the debt crisis, although critics of such attempts believe they are fundamentally flawed because they try to reduce debt burdens while largely leaving free market capitalism and neoliberal structural adjustment intact.

As noted, neoliberal structural adjustment has produced widespread opposition and resistance from scholars, activists and government officials as well as ordinary Latin American citizens who have borne the brunt of such policies. As Woods (2006: 5) writes, opponents assert that IMF and World Bank policies do not work while IMF and World Bank economists say their policies fail if debtor governments do not implement them properly (for such a view see Edwards 2010). What is clear is that the Washington Consensus predictions have often failed to materialize and that the economies of Latin America, despite years of painful adjustment and continued external indebtedness, never ceased to be exposed to crises.

In the early 1990s, Mexico's economic performance was being overwhelmingly endorsed by IMF and World Bank economists as a model of reform and an example for the rest of Latin America. In 1994, the Mexican economy along with the peso collapsed and foreign investors rapidly withdrew from the country; suddenly the IMF was gripped by fears of a "tequila effect" – concerns that the Mexican economic meltdown would affect the rest of the continent. We would witness a similar endorsement followed by meltdown a few years later in Argentina (see Box 3.5).

Solving the debt crisis?

In 1985, US treasury secretary James Baker unveiled the Baker Plan. This involved little more than fresh loans and a call for more structural

Box 3.5

Argentina's economic collapse

Just as the so-called tequila effect in Mexico confounded the neoliberal economists who had expressed satisfaction with the performance of the Mexican economy, Argentina's recent economic history also demonstrates the vulnerability of the neoliberal model. During the 1980s, the Argentine economy was in disarray and so at the end of the decade, after the election of Carlos Menem, Argentina embraced neoliberal reforms. The economy was temporarily stabilized through a strict control of the money supply and a policy that pegged the Argentine peso to the US dollar. In addition, the full IMF recipe was implemented which wiped out local businesses, raised unemployment and increased the costs of newly privatized electricity and telecommunications to users. The government was also running up a large fiscal deficit as a result of the privatization of the state-funded pension scheme, which compromised its ability to maintain commitments to current pensioners. At the same time, however, inflation fell and the economy began to grow again. Consequently, neoliberal economists were pointing to Argentina as evidence of the success of such policies.

However, the benefits were short-lived, and it did not take long for the extreme vulnerability of the Argentine economy to internal and external shocks to become apparent. Over time, the peso became heavily overvalued, capital flight intensified and the economy went into free-fall. Consequently, the government was forced to default on $100 billion worth of debt, freeze bank accounts and devalue the peso. Thousands of Argentines protested angrily in the streets at the situation into which they had been plunged. Many businesses and factories closed, unemployment soared and millions of Argentines fell below the poverty line. Those newly unemployed formed movements of *piqueteros* or Movements of Unemployed Workers (MTDs) and organized blockades and protests. Others became active in the National Movement of Recovered Companies, in which workers who had lost their jobs occupied and took over the failed companies. In many cases, they were turned into successful worker-run cooperatives. High-profile examples include the Hotel Bauen in Buenos Aires and the Zanon tile company in Nequén, which was renamed FaSinPat or Fábrica sin Patrones – Factory without Bosses. There are thousands of Argentine workers who now own and run the factories in which they were previously waged employees.

Argentina's economic situation began to improve with the election of Néstor Kirchner in 2003, who rejected the neoliberal model and began to increase social spending. Within a couple of years, the standard of living improved, the economy began to grow rapidly and millions of new jobs were created. There were concerns that Argentina was heading towards another financial crisis as a result of the global recession in 2008, but although growth rates fell temporarily, a major financial crisis did not take place. Contemporary Argentina is combining global integration with social spending at home, and for now seems to be faring better without harsh neoliberal reforms.

adjustment. It was extremely limited in its aims – indeed it was described by one commentator as the Baker Sham (O'Brien 1991) – but it did recognize that economic growth was failing to materialize. The 1989 Brady Plan, while also calling for yet more structural adjustment, offered a number of mechanisms to convert and reduce debt. Debtor governments could for example buy back their own debt at a discount or convert it into tradable bonds with longer to pay. The plan also created debt-for-equity swaps whereby a foreign investor could buy part of a country's debt at a discount and be paid the equivalent in local currency to invest in that country. Debt-for-nature swaps were also possible, whereby an NGO or environmental organization could buy a country's debt at a discount and would offer full or partial cancellation to the debtor government in return for the creation of a national park. These measures meant that banks had to partially write off some of the outstanding loans (although critics argued that as a result of high interest rates and the flow of dollars from the south to the north the banks had in fact been paid many times over), and some countries were able to achieve limited reductions in their debt burdens this way. The Brady Plan was not a sustainable form of debt relief and Latin Americans continued to be subject to more structural adjustment.

In 1996, the IMF and the World Bank embarked on a more extensive and comprehensive plan to reduce the unsustainable debt burdens of the world's most heavily indebted countries. This plan was called the Heavily Indebted Poor Countries (HIPC) Initiative. The majority of qualifying countries were in sub-Saharan Africa, but five Latin American countries – Bolivia, Haiti, Honduras, Nicaragua and Guyana – were also deemed eligible. In order to qualify for debt relief, debtor governments must have a satisfactory track record in the implementation of economic reforms and prepare a Poverty Reduction Strategy Paper (PRSP) in consultation with civil society. This document has to demonstrate how savings in debt service are to be channelled into poverty-reducing programmes.

Theoretically, the HIPC can provide up to 80 per cent cancellation of a country's external debt. It has two components: the decision point, when debt relief is agreed to, and the completion point, when debt relief is provided. The HIPC was not embraced enthusiastically by campaigners, given the time it took for HIPC countries to get to completion point and also because, like previous attempts to resolve the debt crisis, it demanded more painful structural adjustments. The HIPC does however amount to a recognition on behalf of the institutions of global governance that debt cancellation is the only option, that development does depend on public spending, particularly in areas such as health care and education, and that debt relief and poverty reduction programmes need to be locally generated in a participatory manner in consultation with civil society rather than imposed in a top-down fashion.

The degree to which governments consult with civil society organizations and incorporate their perspectives into PRSPs is of course debatable, but there is a shift in economic thinking at work here. As a result of criticism, the HIPC has become less stringent and debt relief is now being provided more quickly. The HIPC countries in Latin America have managed to reduce their external debts and are now spending more on health care, education and infrastructure than they were in the 1990s. Some additional external debt has been reduced through bilateral debt reduction negotiations and through mechanisms such as the Multilateral Debt Reduction Initiative (MDRI) introduced by the G8 in 2006. In 2003, Nicaragua had one of the highest per capita debts in the world but today its external debt has fallen to $4 billion from a high of $12 billion in 1994 and $6.7 billion in 2003.

Free Trade Agreements

One of the ongoing ways that the thinking behind the Washington Consensus has been expanded in the continent amid increasing contestation has been through the signing of free trade agreements between Latin American countries and the United States. The North American Free Trade Agreement (NAFTA) between Canada, Mexico and the United States came into force on 1 January 1994. NAFTA has involved the gradual elimination of tariffs and non-tariff barriers between the three nations.

Further trade liberalization has had quite contradictory outcomes for the Mexican population. One of the biggest issues is that Mexican-grown maize is being squeezed by US-grown, subsidized and often genetically modified maize, which is having a deleterious effect on Mexican food security and on the livelihoods of those that grow crops for domestic consumption (see Figure 3.4). NAFTA has also led to an increase in the number of maquiladora assembly factories on the Mexican side of the US–Mexican border, providing low-wage employment to hundreds of thousands of Mexicans, mostly young women (see Figure 3.5).

Opposition to NAFTA was central to the Zapatista uprising in Chiapas. On 1 January 1994, the day that NAFTA came into force, thousands of indigenous Mexicans who had formed the Zapatista Army of National

Figure 3.4 *The tortilla, a traditional Mexican staple food made from maize, is threatened by imported US maize*

Source: Marney Brosnan.

Liberation (EZLN) seized a number of towns and cities in southern Mexico to protest at the damage done to livelihoods by neoliberal structural adjustment, which would only be intensified by NAFTA (see Chapters 7 and 8). The US was hoping to get a broad hemispheric agreement with Latin America in place, and had been working for a number of years to enact the Free Trade Area of the Americas (FTAA).

Figure 3.5 *A maquiladora factory in Mexico*

Source: Guldhammer, Wikimedia Commons (released into the public domain).

The FTAA suffered a resounding defeat at the Summit of the Americas in Mar del Plata in 2005, particularly as a result of opposition from Latin American leaders to the maintenance of agricultural subsidies in the US, while the agents of the Washington Consensus have used structural adjustment packages to force Latin American nations to end subsidies.

While the defeat of the FTAA was a big setback for the US, it has proceeded with regional FTAs, such as the Central American Free Trade Agreement (CAFTA), between the United States, Costa Rica, El Salvador, Guatemala, Honduras, Nicaragua and the Dominican Republic. CAFTA provoked widespread social opposition in Costa Rica, and forced the government to hold a closely fought and unequally resourced referendum in which the Yes vote (in favour of CAFTA) achieved a very narrow victory over the No vote (Cupples and Larios 2010; see Figure 3.6). The US has also signed free trade agreements with Chile, Colombia, Peru and Panama.

At the same time, Latin American nations have been producing trade agreements of their own which promote mutual intra-continental trade and economic cooperation and temper the leverage held by the IMF and the World Bank in the continent. Some of these have long histories that

Figure 3.6 *No to CAFTA: a Costa Rican child mobilizes against the Central American Free Trade Agreement*

Source: Julie Cupples.

predate the rise of neoliberalism. Indeed, many Latin American leaders have dreamed of and have been striving for regional integration since the time of Simón Bolívar.

In South America, such agreements include Mercosur between Argentina, Brazil, Paraguay, Uruguay and Venezuela and the Andean Community which comprises Bolivia, Colombia, Ecuador and Peru, with Argentina, Brazil, Chile, Paraguay and Uruguay as associate members. In 2008, Mercosur and the Andean Community merged to form UNASUR. In Central America, the nations have created the Central American Integration System (SICA) which includes Belize, Costa Rica, El Salvador, Guatemala, Honduras, Nicaragua and Panama. In 2004, Venezuela's president, Hugo Chávez, created ALBA, or Bolivarian Alliance for the Peoples of our America, an anti-neoliberal alternative to the FTAA. Initially an agreement between Cuba and Venezuela to exchange petroleum for doctors, it now has eight member states including Bolivia, Ecuador, Nicaragua and three Caribbean nations, Antigua and Barbuda, Dominica, and Saint Vincent and the Grenadines. In defiance of the Washington Consensus and its conditionality, Argentina, Bolivia,

Brazil, Ecuador, Paraguay, Uruguay and Venezuela created the Bank of the South in 2009 with the intention of replacing the IMF and the World Bank as a regional lender. In a somewhat contradictory fashion, Nicaragua is a member of both ALBA and CAFTA. The divergent tendencies evident in the creation of trade agreements both reflect and contribute to the competing regionalizations at work in the continent.

Conclusions

It is now quite widely accepted by analysts that the IMF and World Bank policy prescription, while contributing to the emergence of some improved development indicators, has largely failed Latin America. At times, the institutions of global governance have shown their inability to assess the economic situation adequately, and show little sensitivity to the painful consequences of adjustment for Latin Americans who had nothing to do with the debt crisis. They have, however, had to listen to criticism and respond to it. The discourse of both organizations has shifted dramatically to one of good governance and poverty reduction, although critics assert that this is merely a way of repackaging harsh economic medicine. In the meantime, the economic situation has definitely improved in much of the continent, but at the same time the region remains vulnerable to external shocks.

Neoliberalism is still alive and well in Latin America to varying degrees, although there is no doubt that it has lost its hegemonic status. Some Latin American countries have managed to terminate or substantially change the nature of their relations with the IMF so that they have more control over domestic economic policy. Poverty and inequality remain persistent in the continent, but in recent years Latin America has enjoyed lower rates of inflation and higher rates of economic growth.

Latin America continues to export and depend on commodities. While traditional commodities such as coffee and bananas continue to be important, they have been accompanied by a diversification of the Latin American economy, to include soybeans, ethanol biofuels, flowers, citrus fruits, wine, tourism, and automobile production, not to mention one of most profitable commodities of all – cocaine. Some countries have been able to increase their spending on health care and education. In part, this is because commodities prices have been high. At the start of the millennium, Brazil, with the eighth largest economy in the world, was characterized as a BRIC economy (O'Neill 2001), along with Russia, India and China. BRIC refers to the idea that a major shift is under way in the global economy, whereby economic dominance is shifting to these four countries and away from the G7 countries, which have traditionally had the largest economies.

Brazil continues to export coffee, beef and cocoa, but its sugar is now converted to ethanol biofuels and exported. It also exports orange juice, brans and oils, transport equipment and metallurgical projects. The high price of some of these exports has allowed the Brazilian government to invest in social spending and, under President Lula da Silva, the Brazilian government created a Family Purse (*Bolsa Família*) to subsidize food, school and housing. Families who commit to sending their children to school are entitled to the subsidy. Venezuela has renationalized its oil wealth and is using these resources to fund social programmes.

The struggle for economic justice is a political one. In other words, there is a close relationship between economic equality and political stability. Similarly, the debate over the role to be played by states, markets and social actors in the development process is politically and ideologically driven. The question of development, namely how people are to achieve political freedom and economic dignity, has been a dominant preoccupation across the political spectrum, but the correct approach to achieve such goals is and always has been intensely contested.

Political struggles for development and the attempt to address the persistence of poverty, inequality, exclusion, marginalization and discrimination have taken many forms in Latin America, both violent and peaceful. Economic development is inextricably entangled with politics and ideology. The next chapter will explore the political events, struggles, transitions and transformations of the past century, focusing in particular on the struggles for dignity in the face of terror and economic hardship.

Summary

- After the Wall Street crash of 1929, Latin American countries implemented a development model known as import substitution industrialization. This model had both advantages and disadvantages.

- Given evident failures of modernization, Latin American economists developed dependency theory to explain how the core was causing the underdevelopment of the periphery.

- Changes in the global economy and the increase in the price of oil led to the debt crisis of the 1980s.

- The debt crisis provided an opportunity to implement neoliberal structural adjustment policies in Latin America.

- Structural adjustment had many negative impacts. In many countries, both poverty and inequality increased.

- Structural adjustment is now more contested than ever but continues through mechanisms such as free trade agreements.

Discussion questions

1. What were the causes of the 1980s debt crisis?

2. Who benefits and who is disadvantaged by neoliberal economic globalization?

3. What were the main features of the Washington Consensus in the 1980s and 1990s?

4. Why did the proponents of neoliberal economic policies fail to foresee the collapse of the Mexican and Argentine economies?

5. What are the advantages and disadvantages of free trade agreements for ordinary Latin Americans?

6. How does ALBA differ from NAFTA and CAFTA?

7. What do Brazil, Russia, India and China have in common that has led to the coining of the BRIC acronym?

Further reading

Bulmer-Thomas, V. (2003) *The Economic History of Latin American since Independence.* Cambridge: Cambridge University Press. A good economic history of Latin America.

Franko, P. (2007) *The Puzzle of Latin American Economic Development*, 3rd edn. Lanham, MD: Rowman & Littlefield. A very detailed and informative text on economic development in Latin America, with chapters covering economic policy, trade, poverty, education and health care.

Green, D. (2003) *Silent Revolution: The Rise and Crisis of Market Economics in Latin America*, 2nd edn. London: Monthly Review Press and Latin America Bureau. A little dated given recent events, but an accessible introduction to neoliberal economic policy in Latin America.

Peet, R. (2003) *Unholy Trinity: The IMF, the World Bank and the WTO.* London: Zed Books. While not dealing specifically with Latin America, this is a valuable critical introduction to the three institutions of global governance that have shaped economic policy in the region since the debt crisis.

Potter, G. A. (2000) *Deeper than Debt: Economic Globalisation and the Poor.* London: Latin America Bureau. A text that explains the 1980s debt crisis and its implications for the region and its people.

Films

Commanding Heights: The Battle for the World Economy (2002), directed by William Cran (USA). A documentary series on the global economy including in-depth coverage of neoliberal reform in Latin America.

The Take (2004), directed by Avi Lewis and Naomi Klein (Canada and Argentina). A film that explores the Recovered Factories Movement in Argentina.

The Other America (2010), BBC News and Documentary (UK). A BBC documentary that looks at changing economic conditions in Latin America.
Part 1: www.youtube.com/watch?v=qKFu-OR4B6o
Part 2: www.youtube.com/watch?v=hC4nVzzEhY0

Websites

http://hdr.undp.org/en/reports
Human Development Reports. Produced annually by the United Nations, this report aims to raise awareness of human development. It contains a vast amount of empirical data of use to those interested in Latin American economic development.

http://econ.worldbank.org
World Bank Development Reports. An annual report that has a different theme each year, with a focus on economic development. It includes many development indicators and statistics.

4 Political transitions and transformations

Learning outcomes

By the end of this chapter, the reader should:

- Understand how colonial legacies influence Latin America's political culture
- Be able to evaluate the relationship between the United States and the countries of Latin America and how this relationship has evolved over time
- Be able to discuss the varied political struggles and phenomena that have taken place in Latin America, including revolutions, military dictatorships, the transition to democracy and the "pink tide"

Introduction

Latin America is often unthinkingly characterized by outsiders as a politically unstable region. While political instability is undoubtedly a feature of the region's political landscape – since independence Latin America has experienced civil wars, military coups, dictatorships, IMF riots and guerrilla warfare – it is important to recognize the complexity, heterogeneity and creativity of political struggles and to acknowledge that Latin America has also seen some of the world's most courageous struggles against human rights abuses, most innovative forms of indigenous resurgence and mobilization, and most resourceful political responses to rapid urbanization. The imaginative ways in which many Latin Americans, often against the odds, try to create a better life for themselves and their communities is what keeps scholars of Latin American development making lifelong professional commitments to the political struggles of the continent.

Latin American political culture is both hybrid and dynamic as it draws on diverse European, African and indigenous heritages. The Iberian influence on Latin American politics is widespread and visible. Yet both African and indigenous cultural and political ways of being and knowing continue to intervene to hybridize Iberian ones. Latin Americans, both rich and poor, have also selectively borrowed from other traditions, looking to the United States or to other parts of Europe for political inspiration. At the same time, Latin America has a long tradition of critical intellectual thought rooted in dependency theory, theories of cultural imperialism and internal colonialism, liberation theology and critical pedagogy, which it has exported to other parts of the world (Trigo 2004).

At the current time, Latin America is characterized by some depressing continuities as well as hopeful and optimistic signs of positive change. The failure of neoliberal globalization to enhance people's life opportunities and to bring about sustainable development has led to a decisive rejection of the neoliberal model across the region. Since the late 1990s, the continent has been undergoing what has been referred to as a "pink tide", in which many anti-neoliberal and left-wing governments are coming to power. Far too many Latin Americans still do not get adequate nutrition and health care, live in substandard housing and are unable to finish school. At the same time, public spending on health care and education has increased in some areas, with positive outcomes. Ordinary people are also finding new spaces in which to express their views and exercise democratic citizenship.

The palpable sense that things are improving in the continent is, however, held in check by persistent and growing political challenges posed by the rise in violent crime and drug trafficking and the threat posed by environmental destruction and climate change. These problems have the potential to derail some of the newly won development gains and prevent others from gaining traction.

This chapter charts some of the most salient political struggles and transformations in Latin America since independence. It focuses on colonial legacies, the role played by US foreign policy, the growing political contestation of neoliberal economic policy and the diverse ways in which Latin Americans have responded politically to the question of development.

Colonial legacies

As we saw in Chapter 2, Latin America was brought into being through the violence and brutality of colonialism. The Spaniards and the Portuguese imposed their languages, religion, modes of government, voracious demand for commodities and forms of land tenure on a population they deemed to be inferior. Colonialism brought with it power struggles, cultural conflicts and hierarchies, and its multifaceted legacies have influenced the political trajectories and political culture of the continent in a range of ways. It left three dominant political institutions in place: the landed oligarchy, the Catholic Church and the military, who all continue to exert influence over the development process and everyday politics. We see their influence in the political cultures of religious absolutism, elitism, bureaucratic authoritarianism, *caudillismo*, corporatism, clientelism and populism, which successive generations of Latin Americans across the political spectrum have reproduced, negotiated and resisted in a range of ways. Corporatism can be defined as "the organization of the nation's interest group life under state regulation and control rather than on the basis of freedom of association", while clientelism, or patronage, embodies the idea of giving something in return for a favour (Wiarda and Kline 2011: 13–14).

These phenomena are observed in practices such as civil servants giving state jobs or benefits to family members or political supporters, or when politicians running for office promise improvements in a neighbourhood in return for votes. These practices have also contributed to the creation of large and inefficient state bureaucracies which often encourage corruption and weaken democratic participation (Wiarda and Kline 2011). Military coups – events in which a member of the armed forces seizes power by force – have been common throughout Latin America's history, and indeed the most recent took place in Honduras in 2009 (see Box 4.5).

Populism is also a dominant feature of Latin American political culture and a legacy of the colonial period. Populists are leaders of different political persuasions – they can be left-wing or right-wing – who are able to appeal to and connect with people's everyday lives. They tend to be charismatic leaders with a strong sense of nationalism, and often manage to stay in power for long periods. Many populist leaders combine populism with authoritarianism. Some *caudillos* and dictators are also populists in that they wield a cult of personality and can often enjoy popular support. For example, Juan Perón of Argentina (see Box 4.3) and Gétulio Vargas of Brazil came to power via military coups but once in power introduced popular reforms that resonated with ordinary people. Other populist leaders of the twentieth century include José María Velasco Ibarra of Ecuador, Rómulo Betancourt of Venezuela, Carlos Ibañez of Chile, Carlos Menem of Argentina and Alberto Fujimori of Peru. Between 1934 and 1972, Velasco was elected five times as president of Ecuador (Barton 1997). Populist *caudillos* will often do deals with their political enemies in order to consolidate their longevity in power, as was witnessed in the pact between the right-wing president Arnoldo Alemán and left-wing opposition leader Daniel Ortega, signed in Nicaragua in 1999 (Cupples and Larios 2005).

The struggles for independence did little to undermine Iberian political traditions because they were dominated by white, conservative and often aristocratic creoles. While these actors were opposed to colonial rule, they were of European descent and their aim was to create modern and independent nations like those in Europe. The dominance of Eurocentric thinking meant that even after independence, economic and political conditions did not improve for the poor indigenous, African and mestizo populations. The new independent nations continued to be ruled by elites, and land and other kinds of wealth remained very unevenly distributed, leaving large dispossessed majorities whose living conditions were extremely poor. This state of affairs and the failure to address it adequately in the nineteenth century created prime conditions for social and political unrest and resistance, which have continued up to the present. After independence, colonial legacies were extended and exacerbated by what we might refer to as US imperialism.

US–Latin American relations

A key complicating feature of politics in Latin America, which compounds the multiple legacies of colonialism, is its geographic position to the south of the United States. Since independence from Spain and Portugal, a number of foreign powers, including the British, the French, the Germans and the Dutch, have intervened economically and politically in Latin American affairs, but the relationship with the United States has been the most significant for the region.

In the early 1800s, the United States came to believe that it was its manifest destiny to expand its dominion across the Americas. US Americans began to think of themselves in socially Darwinian terms as superior to Latin Americans (Skidmore and Smith 2005), an attitude that has underpinned a range of foreign policy interventions. In 1823, just as many Latin American nations were gaining their independence from Spain and Portugal, US President James Monroe introduced the Monroe Doctrine which asserted that any attempts by Europeans to further colonize the Americas would be understood as acts of aggression against the United States. In the middle of the nineteenth century, Cuba and

Puerto Rico were described by the US Secretary of State as "natural appendages of the North American continent" and the US tried to purchase Cuba from Spain (Sweig 2009: 4).

Although US foreign policy towards Latin America has taken different forms over the years, it has frequently involved overt and covert political and military intervention designed to protect the economic and geopolitical interests of the US. In 1846, 10 years after the declaration of Texan independence and one year after Texas had joined the Union, the United States invaded Mexico and took one third of its territory. The present-day states of Texas, Kansas, Oklahoma, California, Arizona and Nevada, and parts of New Mexico, Utah, Colorado and Wyoming, had all been part of post-independence Mexico. In 1855, the US administration supported US filibusterer William Walker's invasion of Nicaragua and his attempt to install himself as president and legalize slavery. In 1898, the United States seized Puerto Rico and still occupies this nation today.

The failure to annex Cuba in the same way led the US Congress to pass the Platt Amendment in 1902 in order to curtail Cuban sovereignty. The Platt Amendment stationed US troops in Cuba until 1933, giving the US the right to intervene in the name of "good government". Cuba's first president, Tomás Estrada Palma, leased an area at Guantánamo Bay to the US administration, which was converted into a naval base. The dream of Cuban independence that independence leaders such as José Martí had fought for was thwarted (Sweig 2009). Although the Platt Amendment was repealed in 1933, the US retains control of the naval base at Guantánamo Bay in spite of the fact that it has had no formal diplomatic relations with Cuba since 1959. The controversial base, which since 2002 has been used as a military prison to incarcerate prisoners captured as part of the US War on Terror, is bitterly opposed by the current Cuban government.

Similar controversies surround the treaty signed with Panama in 1903. This treaty gave the US the right to build the Panama Canal and retain sovereign rights over the canal zone until 1999. In 1904 the "Roosevelt Corollary" to the Monroe Doctrine appeared, in which it was made clear that the United States would be prepared to resort to military intervention in Latin America if deemed necessary. It was hardly a measure of last resort. Between the 1846 invasion of Mexico and the present time, the US has intervened militarily throughout Latin America and the Caribbean dozens of times (see Box 4.1 for a selective list). Some countries, especially in Central America and the Caribbean, have been subject to repeated invasions and occupations. Military intervention was often accompanied by "dollar diplomacy". The US sought to increase its

Box 4.1

US military interventions in Latin America

1846	The US invades Mexico and takes one-third of Mexican territory.
1855	An adventurer from Tennessee, William Walker, installs himself as president of Nicaragua.
1894	US marines land in Panama.
1898	US troops occupy Cuba and Puerto Rico.
1903	US troops occupy Panama to gain land for the construction of the Panama Canal.
1904	US troops sent to the Dominican Republic to protect US financial interests.
1905	US troops sent to Honduras.
1908	US troops sent to Panama to intervene in elections.
1910	US Marines occupy Nicaragua.
1912	US Marines occupy Cuba to crush a rebellion by sugarworkers.
1912–25, 1926–33	US Marines occupy Nicaragua.
1915	US Marines occupy Haiti; it becomes a US protectorate until 1934.
1916–24	US Marines occupy the Dominican Republic; it becomes a US protectorate until 1941.
1917	US troops invade Mexico in a failed pursuit of Pancho Villa.
1917–33	US troops occupy Cuba.
1925	US troops sent to Panama.
1932	US sends troops to El Salvador to suppress a suspected communist uprising led by Farabundo Martí.
1934	Nicaraguan nationalist fighter Augusto César Sandino is assassinated with US support.
1946	The US opens the School of the Americas in Panama to train military officers from the Americas in counterinsurgency techniques, including torture.
1950	The US suppresses a movement for Puerto Rican independence.
1954	The CIA overthrows Jacobo Arbenz, democratically elected president of Guatemala, who had begun to implement a land reform that threatened the interests of the United Fruit Company.
1960	Covert actions against Cuba begin.
1961	The US invades the Bay of Pigs in Cuba.
1961	The CIA overthrows José María Velasco Ibarra, the democratically elected president of Ecuador.
1962	The CIA backs a coup to overthrow Juan Bosch, the democratically elected president of the Dominican Republic.
1964	The US backs a coup to overthrow the elected president of Brazil, João Goulart, after he proposes agrarian reform and the nationalization of the oil industry.
1965	The US invades the Dominican Republic to prevent the return of Juan Bosch to power.
1966	The US invades Bolivia to carry out a campaign of counterinsurgency, which culminates with the assassination of Che Guevara in 1967.

1973	The US backs a coup to overthrow the elected president of Chile, Salvador Allende, which brings the Pinochet dictatorship to power.
1980s	The US financially supports the Nicaraguan Contras and Salvadoran and Guatemalan death squads, and creates US military bases in Honduras near the Nicaraguan border.
1983	The US invades the Caribbean island of Grenada.
1984	The CIA mines Nicaragua's harbours.
1989	The US invades Panama to topple President Manuel Noriega, a former CIA ally.
2000	The US provides aid for Plan Colombia to fund counternarcotics operations and military targeting of FARC guerrillas.
2002	The US converts its naval base at Guantánamo Bay in Cuba into a military prison for suspected terror suspects from Iraq and Afghanistan.
2004	The US assists in a coup to oust President Aristide of Haiti.

leverage over Latin American nations by buying up debt owed to other foreign creditors, providing additional loans, seizing customs houses and taking measures to protect US investments there.

US military intervention is motivated by intersecting political and economic interests. The United States has acted to uphold the role played by Latin America during the colonial period as an exporter of raw materials. This model has provided highly profitable economic opportunities for US multinationals such as United Fruit and has ensured a supply of cheap coffee, cotton, sugar and bananas to US consumers. Consequently, there was a vested economic interest in maintaining Latin America's dependent relationship and in thwarting attempts at land reform or self-sufficiency that might have alleviated the misery of the poor majority. In addition, the prime foreign policy goal of the United States during the Cold War was to prevent the countries of Latin America (and elsewhere in the third world) embracing communist ideologies – a goal that turned into an obsession, with tragic consequences.

There were brief moments in the twentieth century when the United States did attempt to establish more reciprocal and cooperative relations with Latin America. In 1933, during the Depression, US President Franklin D. Roosevelt introduced the Good Neighbor Policy, which remained in place until the start of the Cold War in 1945. Although the policy was not entirely altruistic but based on a belief that the United States could better defend its economic interests through non-military means, it did enshrine the principle of non-intervention. It was as a result of the Good Neighbor Policy that the US withdrew its troops from Haiti and Nicaragua and repealed the Platt Amendment.

In 1948, the US and the nations of Latin America created the Organization of American States (OAS), which aimed to promote the mutual security of the Americas and forge beneficial relations between American nations. However, this coincided with the beginning of the Cold War and the concomitant winding down of the Good Neighbor Policy. In 1946, the United States established a military academy, the School of the Americas, in the canal zone of Panama and over the next few decades would controversially train Latin American soldiers in counterinsurgency techniques, including torture (Gill 2004). Its graduates included some of the continent's most notorious human rights abusers, including Leopoldo Galtieri and Roberto Viola of Argentina, Roberto D'Aubuisson of El Salvador and Efrain Ríos Montt of Guatemala. In 1954 the US overthrew Guatemalan president Jorge Arbenz after he initiated an agrarian reform that threatened US economic interests in the region. This act opened one of the saddest chapters in Latin American history. In the next five decades, thousands of Guatemalans would be massacred by their own government, supported by US military aid (see Box 7.1).

Under President J. F. Kennedy, however, there were signs that relations with Latin America might begin to move in a more progressive direction. In 1961, Kennedy unveiled the Alliance for Progress which aimed at economic cooperation and social reform. The Cuban Revolution had triumphed in 1959; the US was determined to prevent another Cuba and recognized the need to improve people's living conditions to prevent them turning to revolution. The Alliance for Progress thus involved a limited increase in aid to the region and encouragement to US entrepreneurs to invest there. The Alliance for Progress was not a great success. Kennedy was assassinated in 1963; his successor, Lyndon B. Johnson, authorized military interventions in the Dominican Republic, Bolivia and Brazil and the US was soon to initiate the war in Vietnam. The Alliance for Progress policy was formally ended by President Nixon in 1969.

With the failing war in Vietnam, counterinsurgency was never far from the mind of US administration officials and its anti-communist obsession intensified. The Cold War might have been a struggle between two opposing superpowers, the United States and the Soviet Union, but in the 1970s and 1980s it was played out in Central America, with devastating consequences for thousands of ordinary people. US military aid also found its way into counterinsurgency campaigns in Colombia carried out against the FARC (Revolutionary Armed Forces of Colombia) and the ELN (Army of National Liberation).

US military intervention in Latin America has declined, in part because the US has been focusing its attention on other parts of the world, namely Iraq and Afghanistan. It does however continue in Colombia, Mexico and other places through the so-called "war on drugs" (see Box 4.6).

Revolutionaries and dictators

US political and military intervention in Latin America is both a contributing factor to and a consequence of the prominence of military dictatorships and revolutionary struggles in the twentieth century. The land tenure system set in place during the colonial period became intensified in the nineteenth and twentieth centuries as a result of the growth of export agriculture driven by both US multinationals and national elite capitalists. This was a system that condemned the large indigenous and mestizo majority to landlessness, hunger and illiteracy, leaving them with little option other than being an exploited labour force on plantations and large farms. It fuelled their sense of political outrage and encouraged political mobilization in favour of political and social change. As a result, the ruling elites frequently resorted to military rule in an attempt to protect the economic status quo and maintain order. Some people also believed that the military were the best equipped to address public sector deficits, protect borders and forge national progress.

In the period prior to the Depression of the 1930s, "order-and-progress dictators" (Wiarda and Kline 2011: 25) often invested heavily in public infrastructure and Latin America's capital cities embraced modern forms of urban planning. The contrast between the upper-class sophistication one found in urban spaces such as Buenos Aires, Caracas and Santiago and the utter misery suffered by the marginalized indigenous and mestizo majority in rural areas further heightened political tensions, especially as the dispossessed majority were deprived of democratic means of expression through which to address their grievances.

The United States also played a key role in the promotion and maintenance of dictatorial rule. By the twentieth century, dictatorship was increasingly seen by the US administration as a more effective and less costly means of controlling its back yard. Direct military control of the Americas was proving complicated for the United States. It was expensive to maintain and produced political opposition within the occupied nations and within the United States. The US administration was in favour of finding new ways to protect its interests. Consequently, the US helped a number of dictators to come to power in Latin America in the 1930s. These included Anastasio Somoza of Nicaragua, Fulgencio Batista of Cuba, Maximiliano Hernández Martínez of El Salvador, Jorge Ubico of Guatemala, Tiburcio Carías Andino of Honduras, Rafael Trujillo of the Dominican Republic and Getúlio Vargas of Brazil. All of these political leaders repressed any political opposition, and did not hesitate to torture, disappear and incarcerate their opponents. And they were all supported by the US administration.

The US Marines occupied Nicaragua almost continuously between 1910 and 1933. This military occupation bred resentment among the local population and fuelled an armed nationalist resistance led by Augusto César Sandino (see Figure 4.2). The United States created a National

---

Box 4.2

The Mexican Revolution

The Mexican Revolution (1910–1920) was the first and one of the most important revolutions in Latin America in the twentieth century. A series of popular uprisings in Mexico had led to the end of the Porfiriato (the dictatorship of Porfirio Díaz). Díaz had come to power in 1876 and ruled until he was ousted by Francisco I. Madero in 1911. Madero had successfully encouraged Mexicans to take up arms against Díaz.

Porfirio Díaz was a typical order-and-progress dictator. Mexico had clearly modernized under his rule, but democracy was elusive and corruption and election rigging were widespread. Industrialists, large landowners and foreign investors had prospered under Díaz, while large numbers of landless *campesinos* lived in economic misery. Madero raised great hopes and contributed to the stirring of revolutionary thought but he mostly disappointed, particularly with respect to land reform.

Madero was assassinated and replaced by Victoriano Huerta, an ally of Díaz, a move that united a number of disparate revolutionary forces, and Huerta too was quickly removed from power. Revolutionary leaders such as Pancho Villa, Álvaro Obregón and Emiliano Zapata (see Figure 4.1) continued to fight for land reform and against the oligarchic nature of Mexican society. The next few years involved bloody and messy struggles and constantly shifting loyalties and rivalries between revolutionary leaders. Land reform, workers' rights and a more limited role for the Catholic Church were enshrined in the new Mexican constitution of 1917 under President Venustiano Carranza. Carranza lost his life before the end of the decade, as did Emiliano Zapata.

In 1920, after the election of Alvaro Obregón, some degree of stability returned to Mexico but many of the goals that the Mexican revolutionaries had fought for would not be implemented until the election of Lázaro Cárdenas in 1934. The new political structures created by the revolution were institutionalized in the Institutional Revolutionary Party (PRI) which ruled Mexico until 2000 – a contradictory political entity known for land reform and nationalization as well as for repression, patronage and electoral fraud.

Figure 4.1 *Photos and images of Emiliano Zapata are on always on sale in Mexico's markets*

Source: Marney Brosnan.

Figure 4.2 *An image of Nicaraguan nationalist fighter Augusto César Sandino in downtown Managua*

Source: Marney Brosnan.

Guard designed to maintain social order within Nicaragua without the deployment of US troops. Anastasio Somoza García, the son of a coffee oligarch, became head of the National Guard and later declared himself president. Sandino was persuaded to come down from the mountains and engage in peace talks, and was duly assassinated by Somoza's guard.

Somoza ruled Nicaragua almost continuously, occasionally appointing other puppet presidents, until he was assassinated in 1956. He was succeeded by his two sons, Luis Somoza Debayle and Anastasio Somoza Debayle. The Somoza family was brutal and corrupt and they enriched themselves at the expense of ordinary Nicaraguans. They tolerated no opposition and became extremely wealthy through embezzlement of state funds, land grabbing, interests in agro-exports such as coffee and cotton, and the granting of concessions to foreign companies. The family enjoyed almost continuous US support until the presidency of Jimmy Carter, who suspended US military aid to Nicaragua.

In Cuba, a combination of political chaos and growing government support for both nationalist and communist policies alarmed both the United States and the sugar plantation owners. This set of circumstances enabled Fulgencio Batista, "a relatively obscure official in the Cuban army of humble origins" (Sweig 2009: 17) to overthrow the government and assume power. Batista, as army Chief of Staff, controlled the presidency of Ramón Grau San Martín and several others. He went on to serve twice as president of Cuba himself, between 1940 and 1944 and again between 1952 and 1959. In 1952, he ran for the presidency but, seeing that his chance of victory was limited, came to power in a military coup with US government support.

During his first presidency, Batista was a reformer and supported progressive policies – legislation in support of labour unions, for example. However, he also worked to protect US economic interests in the region. After 1952, his regime ended all progressive policies. He allowed US oil exploration, protected and promoted US corporations on the island, and turned Havana into a gambling and prostitution mecca for wealthy US Americans. He controlled the government, the media and the education system. Like the Nicaraguans under Somoza, more and more ordinary Cubans were plunged into poverty and miserable living conditions, while Batista and his associates enriched themselves at their expense and the US rich and famous partied in their capital. It is estimated that around 20,000 Cubans were murdered by Batista's forces.

The dictatorial regimes in power in much of Latin America coupled with the misery of the majority meant that socialist and revolutionary thought and struggle flourished in the region. By the time Somoza and Batista had come to power, the Mexican Revolution (see Box 4.2) had already triumphed, putting questions of land reform and social justice firmly on the continental political agenda. Mexico's postrevolutionary condition meant that it became a popular site of exile for those organizing revolutionary movements elsewhere in the continent. For example, the American Popular Revolutionary Alliance (APRA) was created in Mexico by Peruvian socialist leader Victor Raúl Haya de la Torre, and much of the Cuban Revolution was plotted in Mexico by Che Guevara and Fidel Castro. But socialist thought began to flourish throughout the continent and underpinned the creation of political parties and trade unions. Latin American intellectuals such as Peruvians Haya de la Torre and José Carlos Mariátegui began to develop an endogenous socialism embedded in Latin American realities, which became very influential. Some socialist leaders began to pay serious attention to the indigenous question and recognized the importance of indigenous forms of governance and land ownership.

Many of those committed to revolutionary struggle turned to guerrilla warfare. Unable to fight for change through peaceful and democratic

Box 4.3

The contradictory politics of Peronismo

The complexities of military involvement in politics are amply demonstrated in Argentina's political history. From 1932 to 1943, the so-called Infamous Decade, the president of Argentina ruled with military support and supported the Allies during the Second World War. In 1943, there was a coup led by the Grupo de Oficiales Unidos, a group of right-wing officers who wanted to incorporate the political ideas of fascist Italy into Argentina. One of the new leaders was Colonel Juan Perón, who expressed admiration for Mussolini's Italy. As Calvert (2011) notes, in many respects Perón was a dictator, albeit a populist one. The country was a single-party state, he changed the constitution to promote his own longevity in power, and he ruthlessly controlled the media and opposition to his regime. But at the same time, he also improved the condition of the working poor, promoting pro-worker legislation, unionization, land reform, nationalization and poverty reduction through welfare programmes.

Perón's politics were fundamentally contradictory. He promoted socialist and New Deal type policies *and* turned Argentina into a safe haven for Nazis fleeing Germany after the war. Labour organizations were allowed to flourish but were heavily coopted by the state. Perón's wife, Evita Duarte de Perón, who worked to promote social justice and women's rights, was widely loved by the Argentine people. Her premature death from cancer in 1952 created an outpouring of national grief and mourning not unlike that expressed towards Princess Diana in Britain in 1997.

Juan Perón was deposed in a coup in 1955 and was forced into exile in Spain, where he remained until his return to the presidency in 1973. By then, he was too ill to govern and he died in 1974, just two years before Argentina was plunged into one of the most tragic periods in history. Juan and Evita Perón live on in Argentina, as they gave their name to Peronismo, a political movement that underpins the policies of the Justicialist Party, to which both left-leaning (Néstor Kirchner and Cristina Fernández) and neoliberal presidents (Carlos Menem) have belonged. In other words, the politics of Peronismo remain difficult to pin down.

means, they took up arms. For many, it seemed like the only valid political response to the violence and brutality endured on a daily base by ordinary people. In Bolivia in 1952, armed militias who were members of the Revolutionary Nationalist Movement (MNR) overthrew a new right-wing military junta. Led by Victor Paz Estenssoro, the MNR carried out a substantial programme of agrarian reform, increased the provision of education and nationalized the tin mines. The MNR was overthrown by a military junta in 1964. It remained an important political force but it would be another two decades before Paz Estenssoro returned to the presidency. By this time, he had moved sharply to the right and was the president of Bolivia during the neoliberal shock of 1985, discussed in Chapter 3.

The brutality of the Batista regime was the catalyst for the Cuban Revolution, led by Fidel Castro and Ernesto "Che" Guevara. Fidel Castro was jailed for two years for his involvement in an attack on the Moncada army barracks. On his release, he worked with Cubans in Cuba and in exile, organizing the 26th July Movement and deploying troops to the Sierra Maestra. Over the next few years, Castro and Guevara launched a number of attacks on the Batista government. Batista underestimated the threat posed by the guerrilla forces and as a result continued to act repressively towards the Cuban people, which merely had the effect of generating more popular support for the revolution.

The Cuban Revolution triumphed on 1 January 1959, with Batista fleeing the country. Once in power, the new revolutionary government began to implement socialist reforms, introducing agrarian reform and nationalizing US assets. It also began to invest heavily in health care and education, with impressive results. The US, however, was opposed to the Cuban experiment and, fearful that it would export its revolution to other parts of the continent, did everything it could to undermine it. In 1961, it provided support for exiled supporters of Batista to invade Cuba – an attempt that failed. The botched Bay of Pigs invasion was quickly followed by the Cuban missile crisis, in which the US administration discovered that the Cuban government had allowed the Soviet Union to install missiles on its territory only 50 miles from the coast of Florida. The Soviet Union withdrew its missiles in return for a guarantee that the US would not invade Cuba.

While it has not attempted another invasion, in 1959 the US implemented a trade embargo, which eliminated trade between the Cuba and the US (see Figure 4.3), and the CIA has been implicated in a number of assassination attempts on Fidel Castro. Until 1989, Cuba maintained reciprocal trading and diplomatic relations with the Soviet Union, exchanging sugar for oil. US actions have undoubtedly caused hardship on the island (as have the repressive actions taken by the Cuban government against political dissidents) but the Cuban Revolution has hung on for more than 50 years, in some ways against all odds (see Figure 4.4). Che Guevara left Cuba in 1965 in an attempt to export revolution, travelling first to the Congo-Kinshasa and then to Bolivia, where he was assassinated in 1967 by Bolivian state forces in collusion with the CIA.

The Cuban Revolution was a great inspiration to left-wing and progressive forces and to marginalized sectors of society throughout Latin America and beyond. It had the opposite effect on political, military and business elites and the US administration, whose resolve to prevent

Figure 4.3 *Havana, Cuba. The scarcity of vehicle and vehicle parts imports as a result of the trade embargo has kept many old cars on the road*

Source: Marney Brosnan.

Figure 4.4 *References to the Cuban Revolution are ubiquitous in Cuba's urban landscapes*

Source: Marney Brosnan.

any kind of domino effect hardened. The threat of guerrilla warfare and what elite and conservative groups viewed as an urgent need to contain subversives led to the military returning to power in many countries.

In the years following the Cuban Revolution, dictatorships formed all over Latin America. Alfredo Stroessner seized power in a coup in Paraguay in 1954, curtailed civil liberties and, during his 35 years in power, presided over political kidnappings and torture. In Haiti, Papa Doc began his reign of terror in 1957 and employed a secret police force, the Tontons Macoutes, to subdue the Haitian population. In Brazil, two revolutionary groups – the Revolutionary Popular Vanguard (VPR) and the National Liberation Action (ALN) – organized against the military junta that had come to power in the 1960s. The military responded with the brutal arrest and torture of suspected guerrilla sympathizers.

This was a pattern that would be repeated across the Southern Cone, in Uruguay, Argentina and Chile, as well as in parts of Central America and the Caribbean. In Uruguay, a revolutionary movement known as the Tupumaros had gained widespread popular support in the 1960s. The Uruguayan government resorted to increasingly repressive measures in its attempt to contain the Tupamaros, and by 1972 the military had taken control, civil liberties were eliminated and arrest and torture of suspected guerrillas became commonplace. Uruguay became the country with the world's highest percentage of political prisoners, and over the next few decades half a million Uruguayans would be forced to leave the country, among them large numbers of teachers, professors and artists (Salgado 2003).

In 1973, the socialist and democratically elected president of Chile, Salvador Allende, was overthrown in a coup that had the support of the CIA, and a military dictatorship led by General Augusto Pinochet was installed. Supporters of Allende – students, workers and trade unionists – were rounded up in the national stadium and many were killed. Democracy did not return to Chile until 1989, and in that time more than 2000 Chileans were disappeared by state forces.

In Argentina, two left-wing organizations, the Trotskyist Revolutionary Army of the People (ERP) and the pro-Peronist Montoneros, were created in the 1960s to fight for political transformation. By the 1970s, in response to the terror tactics that trade unionists and other left-wing leaders were exposed to by paramilitary death squads such as the Argentine Anti-Communist Alliance, they had added political kidnappings and assassination to their tactics. The response from the military came in 1976, when a military junta seized power in a coup and

Argentina embarked on one of the darkest periods in its history, the so-called Dirty War in which between 10,000 and 30,000 people were disappeared in a state-sponsored reign of terror, dubbed Operation Condor by President General Jorge Videla. All of the Latin American dictatorships of the twentieth century deployed familiar terror tactics such as the use of paramilitary death squads to eliminate "subversives", and many military and paramilitary personnel deployed torture and other counterinsurgency tactics that they had learned at the School of the Americas.

While new dictatorships were being formed in South America, organized and armed opposition against the older ones in Central America was mounting. Honduras, Guatemala, El Salvador and Nicaragua all developed guerrilla armies who formed to try to overthrow the regimes. In the early 1960s in Nicaragua, a political and military organization was created to fight against the Somoza regime. Founded by Carlos Fonseca, Silvio Mayorga and Tomás Borge, the Sandinista Front for National Liberation (FSLN) mobilized militarily in the mountains, cities and universities in an attempt to topple the dictatorship. Many of its members and sympathizers were tortured and imprisoned by the regime, but support for the guerrillas grew. In 1972, a massive earthquake levelled Nicaragua's capital city, Managua, and killed more than 10,000 people. When Somoza embezzled the aid funds, support for the FSLN grew, including from middle-class and conservative sectors of the population. In 1978, 50,000 Nicaraguans demonstrated in the street after the National Guard assassinated Pedro Joaquín Chamorro, editor of the conservative daily newspaper, *La Prensa*. As the regime's brutality grew more indiscriminate, the resolve of the revolutionaries to triumph strengthened. They managed to take all of Nicaragua's towns and cities, taking Managua on 19 July 1979.

The triumph of the revolution initiated a period of social and political transformation. The Sandinista government put emphasis on the mixed economy, non-alignment and political pluralism – principles later enshrined in the Constitution of 1987. Although it never set out to dispossess large and medium-sized agro-exporting landowners, it did implement a series of social and economic redistributive measures aimed at improving conditions for Nicaragua's poor majority. Health care and education were made freely available and people were mobilized on a massive scale in revolutionary health and education programmes as literacy workers or as health *brigadistas*. Vaccination programmes and clean-up days (*jornadas de limpieza*) vastly improved health care, and rates of illiteracy fell significantly, particularly as a result of the 1980

Literacy Crusade when thousands of volunteers travelled to rural areas to teach people to read and write.

The Sandinista government put in motion a process of agrarian reform, which redistributed Somoza's land to small rural producers initially in the form of state farms and cooperatives and subsequently in individual plots. In addition, the government provided credit, agricultural inputs and technical assistance to small farmers to revitalize the farming sector and to attempt to achieve self-sufficiency in basic foodstuffs.

The benefits of the revolution were, however, to be short lived. They soon came under threat after Ronald Reagan became president of the United States and began to fund and equip a counter-revolutionary army known as the Contra, which operated out of military bases in Honduras and Costa Rica. Initially the Contra was composed of former members of Somoza's guard, but as time went on its ranks were increasingly filled by *campesinos* disaffected with the revolution as a result of Sandinista mismanagement and cultural insensitivity.

The demands of the Contra war meant that resources had to be diverted out of social spending and into defence. In 1983, the government introduced military conscription and young men were often forcibly recruited by the Sandinistas to fight the Contra. While between 30,000 and 50,000 Nicaraguans had died in the war against Somoza, a further 30,000 were killed in the Contra war of the 1980s. The schools, health centres and cooperatives that had been created by the revolution became Contra targets and were frequently bombed and destroyed. Many of those killed were teachers, farmers, health workers or environmentalists.

Although the United States never recognized the verdict, US aggression against Nicaragua was condemned by the International Court of Justice in The Hague in 1987. This court also condemned the Iran–Contra scandal, exposed the previous year, in which it was revealed that the Reagan administration had covertly (without the knowledge of Congress) funded the Contras through the illegal sales of arms to Iran. Despite the fact that the FSLN held general elections in 1984 in which it won 67 per cent of the vote, in 1985 the US government imposed a trade embargo on Nicaragua, which caused economic conditions to deteriorate further.

Military rule and guerrilla warfare led by the FMLN (Farabundi Martí Front for National Liberation) were under way in El Salvador. In 1980, the country was shocked when popular bishop of San Salvador, Oscar Romero, who had denounced the human rights abuses, was gunned to death as he gave mass (see Figure 4.5). Subsequently, fraudulent elections brought Roberto d'Aubisson to power and extrajudicial killings and state-

Figure 4.5 *Archbishop Oscar Romero*

Source: Giobanny Ascencio y Raul Lemus- Grupo Cinteupiltzin CENAR El Salvador, Wikimedia Commons
(Creative Commons Attribution-Share Alike 3.0 Unported GFDL (503)2208–3062).

led disappearances became common. The FMLN never triumphed
militarily. In 1992, peace accords were signed, the war came to an end and
the FMLN became a legal political party. It was elected to office in 2009.

By the 1990s, with the end of the Cold War and the collapse of the Soviet
Union, guerrilla struggle based on Marxist–Leninist principles became
increasingly discredited. Revolutionary rhetoric began to give way to a

plurality of discourses grounded in identity politics and diverse grassroots and civil society organizations (see Chapter 6). The guerrillas had triumphed in Cuba and Nicaragua and had gone on to form governments; in other countries they demobilized or signed peace agreements. A number of guerrilla movements continued to be active in the postwar period, and some are still active today.

In 1980, Peruvian Maoist guerrilla group Shining Path or Sendero Luminoso was formed in order to implement a perfect communist system in Peru. Its aim was to convert Peru into a peasant–worker republic in which "the peasantry would be the social basis of revolution" (Barton 1997: 124). Intense conflict between the guerrillas and the Peruvian military, centred particularly in the Ayacucho region, exposed ordinary Peruvians to the terror and atrocities committed by both sides. Many young men suspected of being guerrilla sympathizers were disappeared by the security forces, while the guerrillas killed ordinary villagers they believed were supporting the government. Sendero's violent tactics disgusted many progressive Peruvians. Its strength declined substantially after 1992, when the organization's leader, Abimael Guzmán, was captured, and by 2000 it appeared that the struggle was all but over. The organization is no longer considered a threat to the Peruvian state, but sporadic attacks still occur.

In Mexico in the 1990s, poor indigenous inhabitants of the state of Chiapas formed a guerrilla army, the Zapatista Army of National Liberation (EZLN), and on 1 January 1994 – the day the North American Free Trade Agreement came into force – it seized a number of towns in Chiapas to protest at the extreme hardship caused by further trade liberalization to indigenous economies and ways of life. Although they take their name from Emiliano Zapata (see Figure 4.1), hero of the 1910–1920 Mexican Revolution, the Zapatistas differ greatly from other guerrilla groups in Latin America in that they do not seek to seize state power but rather attempt to enact alternative ways of doing politics. As we shall see in Chapters 7 and 8, the Zapatistas are a fundamental part of the global struggle against neoliberalism and continue to challenge the Mexican state in all kinds of ways.

Two left-wing guerrilla groups active in Colombia's long-standing civil war, the FARC and the ELN, appear to be weakening but continue to launch deadly attacks and kidnappings against state forces and members of their families.

In the twentieth century, the vast majority of Latin American countries succumbed to military dictatorship or guerrilla warfare, or both. Some countries, such as Costa Rica, Ecuador and Venezuela, did enjoy periods of democratic stability and civilian rule in the twentieth century. As a result of the 1948 civil war, Costa Rica abolished the army and established civilian rule. As a consequence, it avoided the human rights abuses, conflict and polarization that plagued its neighbours in Central America and was able to implement public sector investment in education, health care and telecommunications (Cupples and Larios 2010).

Non-military responses to terror

While the political conditions in Latin America made many people feel that they had no option but to resort to armed struggle, there were other forms of political response to the situation in which Latin Americans found themselves in the twentieth century. Most Latin Americans are Catholic, and Latin America developed a specific brand of critical Catholicism known as liberation theology. Latin Americans formed Christian Based Communities (CEBs) and used the Bible to make sense of and respond to the conditions of poverty and suffering in which they were immersed. Peruvian liberation theologist Gustavo Gutierrez (1974) coined the term "preferential option for the poor", which embodies the idea that Christians have an obligation to be concerned about the poor, suffering and marginalized.

Liberation theology was not well supported by the Vatican, which was opposed to what it saw as a problematic association between Christianity and Marxism. The Vatican tried to undermine the movement, but mostly in vain, as many Latin Americans needed to reconcile their faith and their politics and liberation theology provided them with the means to do so.

In Central America, liberation theology was practised by the Jesuits, who spoke out against human rights abuses; as a result, they too became a target of the terror campaign. In 1989, the Salvadoran military murdered six Jesuit scholars along with their housekeeper and her 15-year-old daughter at the UCA, the Jesuit university in San Salvador. So while for much of Latin America's history the Catholic Church has worked to maintain the conservative status quo, in the second half of the twentieth century, a small sector of the Church mobilized politically both to condemn and to try to change the established order. In the past few decades many Latin Americans have abandoned the Catholic Church, while Evangelical churches have experienced exponential growth (see Box 4.4). Religion, politics and development continue to interact in important ways.

Box 4.4

Religion, politics and development

Increasingly, religion is understood as a development issue (see Clarke 2006). Religion influences the political landscape just as politics influences religion, and it does so in both oppressive and empowering ways.

In its long history in Latin America, the Catholic Church has functioned as an instrument of oppression as well as a tool of liberation and empowerment. Since the colonial era, indigenous and African slave populations have blended Catholicism with their own religious and cultural practices to produce creative syncretic forms such as Candomblé, Santería and Umbanda. At times the Catholic Church has defended wealthy elites and at others has come to the defence of the human rights of the poor. In the second half of the twentieth century, liberation theologists combined Catholicism with insights from Marxism to produce the "preferential option for the poor". Local and foreign churches have involved themselves in Latin American development in a range of ways and there are many Christian development organizations, such as World Vision and Christian Aid, that run development programmes in Latin America and have a highly visible presence.

The religious landscape has been in a state of dynamic flux for the past few decades as the influence of the Catholic Church declines and is replaced not only by growing secularization but also by a number of Evangelical churches, especially Pentecostal denominations, with rapidly growing congregations. In Guatemala, Evangelicals make up more than 30 per cent of the population. In Brazil, the somewhat controversial Universal Church of the Kingdom of God, established in the late 1970s, now has more than 5000 churches and owns a number of radio stations and a television channel.

The implications for development and political change of the growth of Evangelical religions are not straightforward, and there is an ongoing debate in the literature about the extent to which evangelical religious participation hinders or foments democratic modes of citizenship and political change. Evangelical religions are often seen as more socially conservative and less socially conscious than many variants of Catholicism, especially as many Evangelical churches encourage passivity and acceptance of suffering and see involvement in politics as sinful. Whereas the Catholic Church has come out in favour of the poor, as it did in Central America in the 1980s, some Evangelical churches supported counterinsurgency movements. Consequently, many development practitioners and activists are dismissive of religion, and see congregations as manipulated by church leaders, distracted from their suffering by joyful clapping and singing.

Believers do however put religion to their own uses. Many Evangelical populations are politically active and see no contradiction between their faith and their political practice. Some non-Catholic and Protestant churches, such as the Moravian Church on Nicaragua's Atlantic Coast, have played important roles in supporting the self-determination of marginalized and indigenous groups. At the same time, there are powerful economic and political interests that underpin the growth of Evangelism. Both the CIA and the religious right in the United States have used religion in Latin America as a means to further their political interests, although not always without resistance from believers.

Rather than make generalizations about the role of religion in development, it is better to theorize religion as a shifting social practice that is promoted and preached by politicians, bishops, priests or missionaries but is also contested, harnessed and used by ordinary people in a range of ways to enact development and fight for dignity and justice. In a text that reflects critically on the growth of Evangelical religions in Latin America, Stoll (1990: xiv) writes that "religion is not just the opiate of the people but their hope for a better world, not just an impediment to social protest but a form of it".

The dramatic growth of Evangelical, especially Pentecostal and Neopentecostal religions, is leading some scholars to ask whether Latin America is ceasing to be a Catholic region. According to Prandi (2008), while the growth of Evangelical religions in countries such as Brazil cannot be denied, they differ substantially from both Catholicism and African Brazilian religions. While Catholic and African buildings, festivals and music are well integrated into Brazilian popular culture and in the process have become secularized, Evangelical religions attract followers but "seem utterly incapable of feeding into Brazil's non-religious culture" (p. 273). Consequently, it might be premature to conclude that Latin America is becoming a Protestant rather than a Catholic continent. Indeed, many Latin Americans that abandon Catholicism do not become Evangelical but abandon religious devotion altogether.

Despite the fear that the repressive atmosphere created, in many countries people overcame their fear and began to mobilize for peace, justice and human rights. As we shall see in Chapter 6, many women, who had never been politically active, began to demand justice. They were usually women whose husbands, sons, daughters or grandchildren had been disappeared or tortured by the state and who were determined to learn what happened to them and bring the perpetrators to justice. They organized as widows, mothers and grandmothers in El Salvador, Guatemala, Argentina and Chile, and had a decisive impact on the Latin American political landscape and changed the way political struggle was understood. Indeed, the political situation was so intense in most parts of Latin America that, as Panizza (2009) notes, it politicized everyday life, and many cultural activities such as sports clubs or music groups became sites of political resistance.

The transition to democracy

By the 1980s, the military governments and dictatorships had become ideologically exhausted and morally discredited. One by one they came to an end and were replaced by more or less democratically elected governments. In 1982, the military junta in Argentina invaded the Falkland Islands/Las Malvinas, a British colony in the Atlantic also claimed by Argentina, in a last-ditch attempt to deflect public attention from the Dirty War. Its defeat 14 weeks later led to the collapse of the government, and an elected civilian government led by Raúl Alfonsín came to power in 1983. Shortly after his election, Argentina's political prisoners would be released. Democratic civilian presidents would be elected across the continent over the next few years; Julio Sanguinetti and José Sarney came to power in 1984 in Uruguay and Brazil respectively. In 1986, Baby Doc would finally be forced out of Haiti. Pinochet and Stroessner both managed to hang onto power until the end of the 1980s. The revolutionary government of Nicaragua, which had been in power since 1979 after overthrowing the Somoza dictatorship, handed power over to a right-wing coalition led by Violeta Chamorro after suffering an electoral defeat in 1990.

New spaces for democratic expression and participation opened up, but in many places democracy remained fragile. One of the reasons was that the transition to democracy coincided with economic neoliberalism, which exacerbated poverty and inequality. In many ways, neoliberalism made it difficult for Latin Americans to reap the benefits of democracy. But the benefits are not to be understated. With the end of the Cold War, the anti-communist hysteria subsided and with it the state-led terror based on the disappearance and torture of suspected subversives, although extrajudicial disappearances and killings are still part of the Latin American political landscape today. Peace accords were signed and truth commissions were set up. Elections that were technically free and fair were held in most countries in the region. In some cases, former guerrilla fighters came to occupy government positions.

Generals also became less likely to resort to military coups, but anti-democratic tendencies have remained. The president of Haiti was deposed in a coup in 2004, as was the president of Honduras in 2009 (see Box 4.5), and there were attempted coups in Venezuela in 1992 and 2002. In the early 1990s, the presidents of both Peru and Guatemala, Alberto Fujimori and Jorge Serrano, committed what were referred to as *autogolpes* or self-coups in which they shut down congress. Fujimori would later be sentenced to 25 years in jail for embezzlement of state funds and use of death squads against Sendero Luminoso. The former president of Nicaragua, Arnoldo Alemán, was convicted of money laundering, embezzlement and corruption and sentenced to a 20 year prison term. In 2009 his conviction was overturned – a move that many Nicaraguans also felt was corrupt.

Box 4.5

The Honduran coup of 2009

The ongoing fragility of democracy was underscored in June 2009 when the democratically elected president of Honduras, Manuel Zelaya, was overthrown in a coup. Soldiers burst into his bedroom and put him on a plane to Costa Rica, and Roberto Micheletti, the speaker of the house, was sworn in as the new president. The military began to clamp down on protests in support of the deposed president. After election, Zelaya had raised the minimum wage and began to align himself with Hugo Chávez's Bolivarian Alternative for the Americas.

Zelaya did not begin his political career as a progressive: he was a member of the Liberal Party and the country's landed oligarchy and political elite. His move to the left, his alliance with Chávez and new-found support for workers and *campesinos* was opposed by members of his party who withdrew their support. The Honduran Constitution forbids re-election and Zelaya wanted to run for another term, so he attempted to hold a referendum on constitutional reform to enable him to do so. His opponents in Congress feared that his re-election would lead to substantial political and economic reforms that would shift the balance of power in Honduras from elite to subordinated groups (Thale 2009). The referendum was scheduled to take place on the day Zelaya was overthrown.

Zelaya was never returned to power. The coup polarized the country, and supporters of Zelaya have continued to mobilize in support of reforms and have continued to be targets of military and police repression.

In Mexico, the political landscape was somewhat different. The PRI, which emerged from the Mexican Revolution, had been in power continuously for 70 years but over that time had moved to the right, had coopted large sectors of Mexican society and was viewed by many Mexicans as a corrupt, incompetent and repressive force. In 2000, after a decade in which Mexico faced both a guerrilla insurgency and financial collapse, the PRI was ousted from power by a right-wing party, the PAN, led by Vicente Fox. Six years later, in an election heavily contested by the left-wing PRD, the PAN returned to power. In this period, drug-related violence has escalated (see Box 4.6). The PRI returned to power in 2012 in elections widely denounced as fraudulent by both the opposition and large sectors of civil society.

The end of the reign of terror and impunity dramatically improved everyday life in the region. Democracy did however struggle to become consolidated in the region, and so scholars coined concepts such as "low-intensity democracy", "schizophrenic democracy", "delegative democracy" and "democratically disguised dictatorships" to describe the

Box 4.6

The "war on drugs"

One of the ongoing ways in which the United States continues to intervene in Latin American affairs is through the so-called "war on drugs". Drug crops are grown and trafficked in many Latin American countries, including Mexico, Colombia, Bolivia and Honduras. In 2000, under President Bill Clinton, the US government launched Plan Colombia, a policy that has involved the aerial spraying of coca fields in Colombia. In a number of countries, US aid became tied to cooperation in eradicating drugs.

In 2007, President Bush launched the Mérida Initiative, which financially underpinned Mexican drug eradication efforts. It was expanded by Barack Obama in 2009 to include "the training of thousands of Mexican agents, the transfer of high-tech weaponry, the deployment of unmanned drones within Mexico and now the direct involvement of Drug Enforcement Administration and CIA agents, U.S. military personnel (from the Pentagon's Northern Command) and private contractors" (Chacon 2011).

Drug trafficking, addiction and drug-related violence are serious problems in Latin America, which are destroying lives and livelihoods and fuelling urban insecurity. The drug trade is estimated to have claimed 50,000 lives during the presidency of Felipe Calderón (2006–2012) in Mexico alone. Guns purchased in the United States frequently find their way into the hands of cartels. In Colombia, the drug trade intensified the militarization of the countryside, as both paramilitaries and guerrillas trafficked and taxed the trade. The magnitude of the problem and the suffering it causes should not be denied. It is however also important to recognize that the lives of many Latin Americans are not touched by the drug trade and that Latin America is often stereotypically and problematically framed as a haven for drug traffickers by outsiders.

The drugs trade also has development dimensions, as drugs money is laundered through legitimate economic activities. At times, drug money is used to fund development initiatives and it does of course circulate in the formal mainstream economy. In the favelas of Rio, for example, drug money has found its way into food and medicine and even the construction of a giant water park built by gang leader Robertinho de Lucas and enjoyed by shanty dwellers (Neate and Platt 2010). On Nicaragua's Atlantic Coast, cocaine wealth leads to the building of homes, churches, schools and health centres and facilitates forms of everyday consumption that would not otherwise be possible given the high levels of unemployment and general lack of economic opportunities that characterize the region (Dennis 2004; Cupples 2012). In the Andean region, many low-income farmers began to grow coca because of the unprofitability of other crops and because of a lack of alternative means of generating income. It brings hardship, violence and fear, but also provides an economic cushion for those living in poverty. Many of the profits from cocaine are, however, enjoyed in the US and Europe, including by first world banks, as drugs money is laundered through the global financial system (Neilson 2004; Vulliamy 2012). Indeed, cocaine is so profitable that any attempt to stop it is futile, as the trade operates according to a balloon effect. As Rosin (2005: 5–6) writes:

> The drug trade, it seems, is more like a balloon than a battlefield. When one part of the balloon is squeezed, its contents are displaced to another. Similarly, when coca

production is suppressed in one area, it quickly pops us somewhere else disregarding national borders. Arrested drug lords are quickly replaced by others who move up the ranks, dismantled cartels are replaced by smaller leaner operations that are harder to detect and deter. When drug-trafficking routes are disrupted by intensive interdiction campaigns, they are simply shifted elsewhere.

The view that the "War on Drugs" led by the United States has not resolved the situation, and may even have made it worse, is gaining traction. According to Livingstone (2009), the policy has been a failure as its only effect has been to harm poor *campesinos* rather than wealthy drug traffickers. Fumigation does not just target coca but also causes great harm to humans, animals, other crops and the environment. It also has not decreased the amount of coca grown, as growers simply move to another area. Given that much of the cocaine trafficked in Latin America ends up in the United States, the US government has also been repeatedly criticized for focusing primarily on supply in Latin America rather than on demand by US consumers and drugs money laundering by US banks.

post-dictatorship condition, which captured the ways in which political participation was coopted by the IMF, the World Bank and national elite politicians (Barton 1997; Panizza 2009). Rising poverty and inequality in urban areas forced more people into the informal sector, where they often struggled to organize politically.

In rural areas, land continues to be an issue. Although the inequality of land ownership has been a pressing political issue throughout the continent since independence, only four countries – Cuba, Nicaragua, Mexico and Venezuela – have ever carried out a substantive agrarian reform policy. In Brazil, landless farmers formed the Brazilian Landless Workers Movement (MST) to address the unequal ownership of land. With 300,000 members, they constitute "the biggest and best organized social movement in Latin America" (Petras and Veltmeyer 2011: 167). They have organized mass occupations of the surplus and uncultivated land of the large landowners. While there is no doubt that they have been successful in getting agrarian reform on the political agenda, Lula's PT government largely failed to deliver on the question of agrarian reform. The ongoing need for agrarian reform has also been central to the Zapatista rebellion since the 1990s.

The end of the dictatorships and the transition to democracy did however coincide with new forms of political organizing. All over the Americas, people mobilized in their communities and around identities (as women, *campesinos*, shantydown dwellers, indigenous peoples, environmentalists and so on) to lobby for land rights, clean water, electricity, garbage collection, public transport, credit, legal aid, education and health care (see

Chapters 6 and 7). Urbanization, secularization and feminism have all been important forces in generating pressure for change. Civil society began to globalize on a greater scale and many Latin American activists have been part of initiatives such as the World Social Forum and the alter-globalization movement more broadly.

The "pink" tide

In many ways, neoliberal economics hindered the consolidation of democracy after the fall of the dictatorships and military governments. But, as Panizza (2009: 24) writes, the economic crisis provoked by neoliberalism "made possible the emergence of new interpretative frameworks that sought to make sense of it and to advance solutions for the reconstruction of social order". In addition, the end of repression made political expression and mobilization easier and by the 1990s, many Latin Americans were becoming increasingly disillusioned with the neoliberal status quo and its inability to deliver prosperity to the majority. This desire for change precipitated a dramatic shift in Latin America's electoral landscape.

With the exception of Chile, Colombia and Suriname, all of the nations of South America were ruled by left-wing and socialist leaders at the start of 2012. The "pink tide" – a term used to refer to a more moderate form of socialism sweeping the continent, which is nonetheless posing substantial challenges to the Washington Consensus – was led by Hugo Chávez in Venezuela. Since coming to power in 1998, Chávez has used Venezuela's oil wealth to fund social programmes. He is a polarizing figure and his calls for twenty-first-century socialism have won him both ardent supporters and detractors.

After the election of Chávez, Latin Americans elsewhere also used the ballot box to press for social and political transformation. Luiz Inácio Lula da Silva, leader of the Brazilian Workers Party (PT), was elected in 2002. In 2010, he handed over power to Dilma Rouseff, also of the PT. Rouseff had been a guerrilla fighter and she was imprisoned and tortured by the Brazilian dictatorship. Left-wing transformation in Argentina has been led by Néstor Kirchner, elected in 2003, and his wife Cristina Fernández who succeeded her husband in office in 2007. Fernández was re-elected as president in 2011.

The left also came to power in Uruguay in 2004, with the Broad Front being elected in both 2004 and 2009. The current president, José Mujica, is a former Tupamaros guerrilla fighter. In 2006, Evo Morales, an Aymara trade union leader and coca grower and leader of the Movement for Socialism (MAS), was elected president of Bolivia. 2006 also saw the election of Christian socialist Rafael Correa in Ecuador. In 2008, former bishop and liberation theologist Fernando Lugo and his newly created and socialist-oriented Patriotic Alliance for Change were elected to power in Paraguay. In 2012, Lugo was impeached by congress and removed from office. In 2011, socialist Ollanta Humalla became president of Peru. Chileans elected centre left presidents in both 2000 (Ricardo Lagos) and 2006 (Michelle Bachelet), but the right was returned to power in 2010.

Similar moves have taken place in Central America. In 2006, former FSLN guerrilla leader and revolutionary president Daniel Ortega was returned to power after 16 years in opposition and he was re-elected (somewhat controversially, as re-election is not permitted according to the Nicaraguan Constitution) in 2011. Both El Salvador and Guatemala have elected left-leaning presidents, although the right was returned to power in Guatemala after just one term. In El Salvador, the president Mauricio Funes is a member of the FMLN, the former guerrilla organization that became a political party after the peace accords of 1992. He did not fight in the civil war but had worked as a journalist and as a result came into contact with many of the ideas of the FMLN. In Honduras, President Manuel Zelaya moved to the left after election and ended up losing the support of his party (see Box 4.5).

Conclusions

Latin America's political landscape has produced a diverse range of political actors – including military dictators, authoritarian populists, revolutionary guerrilla fighters, politicized mothers, mobilized *campesinos* and resistant shanty town dwellers – who have all been engaged in the struggle for hegemony. These struggles and the shifts and transformations they have produced underscore the intense contestations that surround how to do development in Latin America. Many Latin Americans have lost their lives fighting for development and social justice. Dealing with the legacies of these deaths – indeed, ensuring that people did not die in vain – poses ongoing challenges for the development project in Latin America. Political sympathies are often powerfully felt and political leaders are both admired and reviled, depending on who you talk to.

After the desperation of the lost decade, there is now a sense of political optimism, although not experienced or shared by all. The political changes under way in Latin America are significant for the region and the world. It is important to stress that while a trend towards socialism appears to be in place and the changes in the electoral landscape constitute a decisive challenge to the legitimacy of the Washington Consensus, the parties and leaders in power are actually quite heterogeneous and many of them display the populist, personalizing and *caudillista* tendencies of earlier generations of political leaders. Some are well connected with the popular sectors; others less so. Some of them, such as Daniel Ortega, appear to act in quite contradictory ways. Nicaragua became part of the Bolivarian Alternative for the Americas (ALBA), but did not pull out of CAFTA and the Nicaraguan government continues to sign agreements with the IMF. Similarly, in Brazil Lula followed a much more moderate course than many had hoped. There is no doubt that neoliberalism continues in the region but in a much more hybrid and less dogmatic way than in the 1980s and 1990s.

At the same time, there is clear evidence that these changes are promoting more participatory forms of democracy, there is more social investment which is reaching the excluded sectors of the population and, as Wiarda and Kline (2011) note, Latin America does appear to be shaking off its corporatist past. Large municipalities in cities such as Porto Alegre and São Paulo are experimenting with some degree of success with innovative development models such as participatory planning and budgeting, giving ordinary people a voice in political decision making (see for example Krantz 2003). In addition, Latin American militaries are now smaller and more professional and the Catholic Church, once a decisive political actor, finds its influence diminished, challenged not only by left-wing thought but also by urbanization, feminism, growing secularism and Protestanism. Marginalized sectors of the population have gained new spaces for expression and agency both within and beyond the formal political sphere.

Chapters 6 and 7 develop these arguments, showing how identities of "race", gender, sexuality and class can be both forces of oppression and sources of mobilization.

Summary

- Latin America's political culture has been characterized by elitism, corporatism, authoritarianism, clientelism and populism, legacies of Iberian colonial rule.

- Latin American development has been complicated by US imperialism. The US has intervened militarily in Latin American countries many times since independence.

- Latin America has experienced both military dictatorship and revolutionary struggles.

- Mexico, Cuba and Nicaragua had successful revolutions.

- Torture, disappearances and extrajudicial killings were common during the military dictatorships of South America.

- Religion and development interact in important ways. While Catholicism has long been the dominant religion, many Latin Americans have joined Evangelical churches.

- In the 1980s and 1990s, most of the dictatorships fell and democratically elected governments came to power.

- In more recent years, many left-wing governments have been elected on anti-neoliberal platforms, a phenomenon dubbed the "pink tide".

Discussion questions

1. Why was twentieth-century Latin America characterized by political instability in the form of both dictatorships and revolutionary warfare?

2. Why did Latin American governments act with such brutality against their own people in the 1970s and 1980s?

3. Land reform is no longer the pressing issue it once was. Discuss.

4. Explain what is meant by twenty-first-century socialism. What are the distinguishing features of the contemporary political landscape in Latin America?

5. Some people adore him and others loathe him. Why do you think Venezuela's Hugo Chávez is such a polarizing figure in Venezuela and beyond?

Further reading

Barton, J. R. (1997) *A Political Geography of Latin America.* London: Routledge. An introduction to political development in Latin America from a geographical perspective.

Burdick, J., Oxhorn, P. and Roberts, K. M. (eds) (2009) *Beyond Neoliberalism in*

Latin America. New York: Palgrave Macmillan. An edited collection that deals with how Latin Americans are imagining and forging alternatives to neoliberalism.

Grugel, J. and Riggirozzi, P. (eds) (2009) *Governance after Neoliberalism in Latin America.* New York: Palgrave Macmillan. An edited collection that explores post-neoliberal forms of state governance and activism and their implications for development and democracy.

Panizza, F. (2009) *Contemporary Latin America: Development and Democracy beyond the Washington Consensus.* London: Zed Books. An accessible introduction to political development in Latin America over the past three decades.

Films

Missing (1982), directed by Costa-Gavras (USA). Starring Jack Lemmon and Sissy Spacek, this film delves into post-coup Chile in 1973 through a US businessman who travels to Chile in search of his missing son.

Death and the Maiden (1994), directed by Roman Polanski (USA/UK/France). Based on a play by Ariel Dorfman, a woman meets the man she believes tortured and raped her during the military dictatorship, and takes him captive in her home in search of a confession.

Carla's Song (1996), directed by Ken Loach (UK). A film set in 1980s Glasgow and Nicaragua in which a Glaswegian bus driver meets Carla, an exiled Nicaraguan living in Glasgow, and travels with her to Nicaragua in the thick of the Contra War.

Evita (1996), directed by Alan Parker (USA). Controversially starring Madonna as Evita Perón, this is a musical dealing with this period in Argentina's history.

The Revolution Will Not Be Televised (2003), directed by Kim Bartley and Donnacha Ó Briain (Ireland). An Irish film crew is making a film about Hugo Chávez when he is temporarily ousted in a coup.

War on Democracy (2007), directed by Christopher Martin and John Pilger (UK). A documentary that explores US interests and intervention in Latin America and the US role in overthrowing democratically elected presidents and supporting military dictatorships.

Red Oil (2008), directed by Lucinda Broadbent (UK). A documentary that explores the controversial use of oil wealth to fund social programmes in Venezuela under Hugo Chávez.

Che (2008), directed by Steven Soderbergh (USA). In two parts, this is a film that documents the revolutionary struggles of Ernesto Che Guevara. Part 1 shows the toppling of the Batista regime in Cuba, Part 2 Che's campaign in Bolivia.

South of the Border (2009), directed by Oliver Stone (USA). A tour around South America and Cuba to meet some of the left-wing presidents associated with Latin America's "pink tide". Stone interviews Hugh Chávez, Evo Morales, Rafael Correa, Néstor Kirchner, Cristina Fernández, Lula, Fernando Lugo and Raúl Castro. A sympathetic treatment of the "pink tide".

Codename: Butterflies (2009), directed by Cecilia Dominguez (USA). A documentary that tells the story of the Mirabel sisters and their courageous struggle against the Trujillo dictatorship in the Dominican Republic.

5 Latin America's environments

The struggle for sustainable development

Learning outcomes

By the end of this chapter, the reader should:

- **Be able to explain the relationship between colonialism, neoliberalism and environmental degradation**
- **Understand some of the main environmental challenges facing cities and rural areas and the diverse responses to them**
- **Understand why environmental issues such as deforestation and climate change could be conceptualized instead as development issues**

Introduction

Latin America was brought into the global geographical imaginary as a site to be ravaged and exploited by outsiders. Since the beginning of the colonial period, Latin America's bountiful natural resources have been highly sought after. Mining and agriculture were pre-Columbian activities and before the conquest, the indigenous inhabitants grew maize, beans, tomatoes, potatoes, chilli peppers, squash and avocados and they made jewellery and other items out of gold and silver. But the ways in which crops were cultivated and minerals extracted changed dramatically with the arrival of the European colonizers.

Colonization was built on the rapid plundering of alluvial deposits of gold and silver, followed by mining and plantation agriculture and the creation of an unequal system of land tenure that concentrated land, resources and wealth in the hands of a minority. The colonizers also brought new crops to the continent, such as sugar, coffee, wheat and citrus fruits, which led to profound land use transformation. Mining and plantation agriculture often led to the displacement of small and subsistence farmers, condemning them to landlessness and hunger and converting them into an exploited labour force. The obsession with gold and silver and then with "King Sugar" would be followed by other equally destructive "agricultural monarchs" – rubber, chocolate, coffee, bananas (see Box 5.1 and Figure 5.1), beef, wheat and minerals such as nitrates, copper, tin, iron ore, bauxite and titanium (Galeano 1971).

The colonial period initiated a process of large-scale ecological and social transformation that has continued to the present day. Such unsustainable practices proliferated after independence as US and other foreign investors and corporations also moved into the region to exploit its lands, soils, crops, forests, seafood and minerals along with human labour for profit. The extractive economy brought vast wealth and prosperity for a minority, but misery for the majority and devastation for the environment. Many countries and regions would become dependent on a single crop or mineral, and given that they were all subject to boom-and-bust cycles, economic misery would result when the price plummeted, as it often did.

This economic model was intensified in the wake of the 1980s debt crisis, which encouraged resource extraction and export as a means to raise foreign exchange to service debt. As Franko (2007) notes, the environmental costs of such a model have been substantial. While Latin American economists and dependency theorists have long argued for the need to shake off a development model based on the extraction, cultivation and export of primary products, neoliberal structural adjustment has impeded this endeavour.

In addition to dealing with the environmental challenges posed by centuries of commodity trade, Latin America faces additional challenges of a tectonic and climatological nature and much of the continent can be described as hazard-prone. It has faced devastating earthquakes, hurricanes, landslides and floods, many of which have led to substantial loss of life and livelihood destruction. It is important, however, not to simply blame nature for such losses. As we shall see, Latin America's hazard susceptibility is exacerbated by development models in place. It is important to recognize that for many Latin Americans, rural and urban, both environmental degradation and exposure to environmental hazards are far more immediate and serious than they are for many first world inhabitants. Rather than understanding local natural environments as sites of wilderness, recreation or relaxation, for many Latin Americans these environments are central to physical and cultural survival. Environmental struggles are "usually about basic needs, cultural identity and struggles to survive, rather than about providing a safety valve within increasingly contested urban space" (Goodman and Redclift 1991: 4). When these environments become degraded through drought, pesticide contamination or other hazards and people can no longer grow the subsistence crops

Box 5.1

Bananas

In many ways the banana epitomizes everything that is socially and environmentally unsustainable about Latin American development. The banana is the commodity that has had the most transformative effect on the ecological, social and cultural landscapes of Latin America, and the banana trade was built on various kinds of abuses (Moberg and Striffler 2003). Bananas are not native to Latin America but have been cultivated there since the colonial period. In the late nineteenth and early twentieth centuries they began to be consumed by people in Europe and the United States, and are now the most widely eaten fruit in the world.

The banana has become synonymous with political dependency and economic exploitation. The term "banana republic" was coined to refer to a country such as Honduras or Guatemala in which multinational banana companies could dictate domestic policy. US multinationals the United Fruit Company (UFCO, now called Chiquita) and Standard Fruit moved into Central America, Colombia and parts of the Caribbean in the early twentieth century and soon commanded countless acres of fertile land. UFCO became the largest employer in Central America, and gained a reputation for intervening in domestic policy in order to protect its interests. The president of Guatemala, Jacobo Arbenz, was overthrown by the CIA when he threatened to confiscate 15 per cent of UFCO's 650,000 acres in an agrarian reform programme and enhance worker protection legislation. While multinational banana companies are often presented as all-powerful manipulators of local people, as Soluri (2005) remarks, a growing body of historical scholarship demonstrates the ways in which the fruit companies' activities were challenged by locals in Central and South America, through diverse modes of negotiation and popular resistance including strikes, public protests and squatter settlements (see also Euraque 1996; LeGrand 1998).

The environmental effects of large-scale banana cultivation have also been devastating. Bananas grow well in Latin America as part of diverse ecosystems (see Figure 5.1), but the multinational approach has been to create vast plantations – essentially a monoculture – which tends to have a negative effect on local biodiversity. Monoculture also makes crops less resistant to disease, and Latin American banana plantations have been seriously afflicted by plant epidemics such as Panama disease and Sigatoka. Vast areas of land were deforested to make way for plantations, and waterways were harnessed to irrigate them. In addition, bananas grown in this way reduce the fertility of the soil and so require large quantities of pesticides and insecticides, which find their way into local soils, waterways and human bodies. US multinational fruit company Dole, for example, continued to use the pesticide Nemagon produced and exported by Dow Chemical and Shell in its banana plantations even after it was banned in the United States. The use of Nemagon has led to dramatically high levels of sterility, miscarriages, birth defects, cancer and skin diseases among banana workers and their families. The workers, many of whom were very sick from pesticide exposure, held continuous street protests to demand justice in Managua for many years (Guevara 2008).

Figure 5.1 *A mule transports bananas grown on a small family farm in Nicaragua*

Source: Marney Brosnan.

they need to survive, they are often forced to migrate. Effectively, many Latin Americans have become environmental refugees.

This chapter explores the environmental dimensions of Latin American development. After an overview of the relationship between neoliberalism and environmentalism, it focuses on forests, mining and oil. It then outlines Latin America's susceptibility to hazards and disasters, urban environmental issues and the question of climate change. Finally, it considers some of the ways in which Latin Americans are struggling to promote more sustainable modes of development.

Neoliberalizing the environment

One of the ironies of Latin American development after centuries of pillage and ecological destruction was that neoliberal economic policies were introduced precisely at a time of growing global and continental awareness of environmental issues. Neoliberal economic policies, with their emphasis on export-oriented growth and free trade agreements, have however meant an intensification of environmentally destructive extractive industries and plantation agriculture. Consequently, an opportunity to harness growing environmental concerns and begin to implement more sustainable modes of development was lost, with tragic effects for both humans and nonhumans.

Large agricultural operations can often have a very negative effect on local food security and can exacerbate malnutrition, as they result in crops being grown for export rather than domestic consumption. Since the debt crisis, Latin America has continued to produce traditional exports such as bananas and coffee (see Figure 5.2), and mining operations are currently expanding all over the continent. Economic globalization is producing new patterns of investment in non-traditional

Figure 5.2 *Coffee beans grown in Nicaragua*

Source: Marney Brosnan.

agricultural exports (NTAEs), such as broccoli, citrus fruits, flowers and melons. In some cases, they provide welcome agricultural and economic diversification, disrupting dependency on a single crop or mineral, but they also often involve the heavy use of pesticides, insecticides and fertilizers and the further concentration of land and wealth. Many pesticides are toxic to humans; some are known carcinogens and have been responsible for the ill health and premature death of farm labourers. Toxins also find their way into soils and streams.

Cattle production has also increased to meet global demand for meat, including fast food and pet food, and many crops, such as sugar and soybeans, are now being grown to produce biofuels – largely to fuel cars in the first world rather than to feed people in the third world.

Latin America's environmental refugees, fleeing land degradation or displaced from their land by large agribusiness, often become exposed to new risks after migration. All too frequently, they relocate to an area that poses its own environmental risks. They might for example be forced to move into the rainforest where they clear trees in order to grow crops, or they might settle on hazardous land, such as on the slopes of hills or volcanoes or on riverbanks, where they are prone to landslides or flooding. The combination of environmental degradation, livelihood constraints and economic policy has also led many rural inhabitants to migrate to cities, where they often illegally squat on unused land.

One of the dimensions of neoliberalism that many Latin Americans find disturbing is that privatization is now being extended from state-owned enterprises and national industries into natural resources such as water, fisheries, forests and biodiversity. The privatization of water in Latin America began in Chile with the 1981 Water Code, which was touted as a great success by the World Bank (Liverman and Vilas 2006). Subsequently, Mexico and Argentina embarked on water privatization programmes. It is important to note that the public water systems that the privatized systems replaced were often deemed inefficient. By the end of the "lost decade", most public water systems were in urgent need of infrastructure improvement. It was felt that private water companies would have the resources to make improvements to the system that the cash-strapped governments lacked. But given that no human being can survive without water, many people are opposed to its privatization and consider water to be something that belongs to the commons and that cannot therefore be commodified. Water privatization (see Figure 5.3) became a high-profile political issue at the end of the 1990s when the World Bank informed the Bolivian government that it would not

Figure 5.3 *A water truck in Puebla, Mexico; selling purified water in Latin America, where the safety of drinking water cannot be guaranteed, is a lucrative business*

Source: Marney Brosnan.

guarantee a loan unless the water supply of the city of Cochabamba, Bolivia's third largest city, was sold to the private sector. This move would lead to the Cochabamba Water War of 2000 (see Box 5.2).

The neoliberal mandate to attract foreign direct investment and to cut public spending works against the creation and enforcement of environmental protection legislation. In order to attract multinational investment, governments feel pressured to relax environmental demands on companies. The creation of free trade agreements has made environmental enforcement even harder. As Liverman and Vilas (2006: 337) note, NAFTA contains protections for investors against the taking of their property to enable them to maximize their investment, a mechanism that makes it difficult for local actors to protect their local environments. They refer to how a US company, Metalclad, was prevented from establishing a landfill company in Mexico because the municipality had fears about water pollution, and as a result the company claimed $15 million in compensation.

In areas where there are large numbers of maquiladora factories, such as Ciudad Juárez and Tijuana in Mexico, toxic environmental pollution from textile, television and toy factories is impacting on the health of residents. While not all maquiladoras are environmental polluters, many are and the factories are responsible for producing several million tonnes of hazardous wastes, most of which is dumped in Mexico (Roberts and Thanos 2003). As Shiva (2002) states, safe drinking water is so scarce in the maquiladora region of Mexico that babies are often given Coca-Cola to drink instead of water.

Box 5.2

The Cochabamba Water War

Prior to privatization, public water provision in Cochabamba was inadequate. Residents had therefore developed their own water harvesting strategies through the digging of communal wells or the installation of roof tanks to collect rainwater. As part of a World Bank-mandated privatization process, the Bolivian government leased the water supply of Cochabamba to a US company, Bechtel, in a deal negotiated in secret. After taking over the water supply, Bechtel hiked water rates by 200% and residents were told that they were required to have a permit to take water from their own wells and tanks.

The people of Cochabamba organized street protests on a massive scale and for three days Cochabamba was paralysed in a city-wide general strike. The Bolivian military violently repressed the protests but the people refused to back down. Local organizations made up of *campesinos* and indigenous people produced a counterproposal that rejected the commodification of water and moved to define water as "a social and ecological good that guarantees the well-being of the family and the collectivity and their social and economic development" (cited in Assies 2003: 16–17). In the end, Bechtel was forced to flee Bolivia and the water supply was re-nationalized. The water war had been fought and won by a diverse coalition of political actors, who had mobilized alternative understandings of water based not on individual ownership but on collective Andean cosmologies (see Laurie, Andolina and Radcliffe 2002; Assies 2003; Perreault 2005).

Forests

Most people outside of Latin America would probably identify the deforestation of the Amazon rainforest as the most serious and salient environmental issue in Latin America. In the 1980s, the Amazon rainforest emerged as a threatened habitat not just for its human and nonhuman inhabitants but for the entire planet. Often referred to as the "lungs of the world", the Amazon rainforest came to symbolize a planet in peril. As Milton (1993) states, environmental problems are now perceived to be global in scope and we have become accustomed to framing the environment in global terms. But it was not always so: it was concerns surrounding the deforestation of the Amazon that led to the globalization of environmentalism. Today the Amazon is an imbroglio of partially enforced management plans and legislative frameworks, land grabbing and speculation, illegal logging, human rights abuses and widespread ecological destruction.

Rapid deforestation of the Amazon did not begin until the 1970s. There were no roads in the region until the Belém–Brasília highway was constructed in 1958. According to Fearnside (2005), five centuries of European presence prior to 1970 resulted in a deforested area only slightly larger than Portugal. Twice this amount was lost in the decade from 1980 to 1990 (Kaimowitz, Mertens, Wunder and Pacheco 2004). At the present time, the deforestation of the Amazon is estimated to be occurring at the rate of an area equivalent to 11 football fields every minute (Nordhaus and Shellenberger 2007).

A series of key events in the 1980s intensified global awareness of the plight of the Amazon rainforest. Initially, activists and development practitioners became concerned by the negative environmental impacts that World Bank projects were having in the Amazon. For example, the Carajas iron-ore project in the Brazilian Amazon was funded by the World Bank in 1982 and designed to promote economic growth in the eastern Amazon. Lack of concern for the social and environmental dimensions of the project contributed to the acceleration of deforestation and other forms of environmental degradation. It also contributed to a sharp increase in land values, which displaced low-income farmers and indigenous peoples from the region (Oxfam Policy Department 1995). Similarly disastrous outcomes were associated with the Polonoreste project, a World Bank road building and colonization project initiated in 1985 in the Brazilian province of Rondônia. The project promised land to landless people in an area already inhabited by around 40 different indigenous groups. As Hoy (1998: 59) notes:

> The results, even by World Bank standards, were disastrous. By 1987, almost all of the jungle had been slashed and burned by land clearers, whose crops had scarcely a chance of producing anything amidst the ash and burned soil. Satellite photos taken that year showed six thousand forest fires burning across the entire Amazon basin – the single largest human-caused change on earth readily visible from space. The incidence of malaria approached 100 per cent in some areas, sending failed settlers back to their urban centers carrying a highly resistant and lethal form of the disease. The original inhabitants, the tribal groups, fared even worse; many of them were menaced with physical extermination by measles and flu epidemics.

US-based environmentalists, concerned by the environmental impact of World Bank loans, focused their campaigns on the rubber tappers in the state of Acre who were fighting to defend their land-use rights. As Keck (1995) writes, they developed a project to protect their livelihood which was brought by US activists to the international community. It was supported by the multilateral development banks, and consequently the rubber tappers led by union leader Chico Mendes gained "a voice in political arenas they could not have reached alone" (Keck 1995: 39)

Chico Mendes was not really an environmentalist. His struggle was about securing access to health care, education and employment. These aims did however require the protection of the rubber and Brazil nut trees under threat by ranchers, which produced an alliance with Northern environmentalists concerned with conservation and sustainability (Revkin 2004; Nordhaus and Shellenberger 2007). At the end of the 1980s, international interest in the Amazon exploded, in part as a result of the release of photographic, video and satellite imagery that showed unprecedented burning in the Amazon (see Figure 5.4). On one day in 1987, the NOAA-9 satellite captured more than 7000 fires (Humphreys 1996). In 1988, an image taken from a space shuttle flight showed a smoke cloud the size of India from burning in the Amazon rainforest (Revkin 2004). Global awareness was further strengthened in December 1988 when Chico Mendes was assassinated and his death made front-page international headlines.

This global concern notwithstanding, the deforestation of the Amazon rainforest and other tropical forests in Latin America continues to

Figure 5.4 A satellite image taken in 2007 shows fires and deforestation on the Amazon Frontier, Rondônia, Brazil: such images were instrumental in boosting awareness of damage to the rainforest

Source: NASA Earth Observatory, Wikimedia Commons (released into the public domain).

advance rapidly. Between 1980 and 1995, at least 3 million hectares of Amazon rainforest per year were lost (Stea and Lewis 2011: 59). The biggest contributors to deforestation in the Amazon are cattle ranching, logging for tropical hardwoods and biofuel production. According to statistics produced by Greenpeace (2009), the number of cattle in Brazil more than doubled between 1990 and 2003 from 26.6 million to 64 million.

Although the focus of global campaigning on deforestation has been on the Amazon rainforest, the pace of deforestation is actually faster in Central America (Stea and Lewis 2011: 59), also fuelled by the growth of large agribusiness, the conversion of agricultural land to pasture, commercial timber extraction and the landlessness of the poor majority. Landless *campesinos*, environmental refugees from soil erosion or people who have lost their land to agroexports sometimes expand the agricultural frontier by moving into tropical rainforests where they clear trees in order to plant subsistence crops. Rainforest soils are not suitable for sustained agriculture and once stripped of trees, the land rapidly loses its fertility, the nutrients are washed from the soil and the family is forced to move further into the forest (Carriere 1991).

Green (2006) outlines the substantial costs of deforestation on climates, soils and people. Deforestation disrupts the local climate since trees regulate the storage and release of rainwater. Forests also stabilize soil, so deforestation rapidly leads to soil erosion and other forms of land degradation. Droughts, floods and landslides are more likely in deforested areas, putting lives at risk. Deforestation also disrupts global climate. Tropical forests are carbon sinks, while forest fires and deforestation release substantial greenhouse gases into the atmosphere. Rainforests are also a vital source of new genetic and pharmaceutical material. The destruction of the Amazon and other rainforests might therefore impede our ability to find cures for serious diseases such as cancer or HIV/AIDS.

Consequently, the Amazon rainforest is widely seen as a key provider of global environmental services including biodiversity maintenance, carbon storage and water cycling (Fearnside 2005), a view that sometimes provokes a nationalist backlash from Brazilians who wonder why Europeans and US Americans who developed by destroying their own forests put so much pressure on them not to do the same (Nordhaus and Shellenberger 2007). Indeed, one of the problems identified with the transnational activism to save the Amazon has been its focus on deforestation as an environmental issue rather than a broader development issue. It is important to recognize that the deforestation of the Amazon is driven by other development issues facing Brazil, including poverty, unequal land tenure and distribution of wealth, the lack of public investment in education and health care during the 1980s and 1990s, the servicing of Brazil's external debt, and the everyday violence suffered by street children (see Nordhaus and Shellenberger 2007). National and global initiatives to halt deforestation such as Payment for Ecosystem Services (PES), where landowners are paid *not* to deforest their land, and the UN-led Reducing Emissions from Deforestation and Degradation (REDD) and REDD+, an extension of REDD to include conservation and enhancement of carbon stocks, are extremely controversial and at best are producing mixed results. While they have the potential to promote sustainable forest management, they function by subjecting natural resources to market logics – viewed by some as a kind of greenwashing of capitalist activities. Indeed, they often result in payments to wealthy landowners or they pay out for forests that were not threatened anyway (Kaimowitz 2008). In addition, detractors argue that such schemes allow first world polluters to continue to generate emissions by paying third world forest managers (see redd-monitor.org 2011).

Mining

In the neoliberal and post-neoliberal era, mining has been an important economic activity in Latin America. The high prices generated by the recent global mineral boom and demand for metals from emerging economies in Asia encourage substantial investment by foreign companies in mining operations. As Bridge (2004) notes, many nations have revised their mining codes or laws in order to attract foreign investment and Latin American nations are among them. Several US and Canadian companies, including Barrick Gold, Goldcrop, Meridian, Phelps-Dodge and the Newmont Mining Corporation, are running large-scale mines, sometimes in operation with local partners. Gold, silver and copper are the most sought after minerals, along with Latin America's remaining reserves of nickel, lead, zinc and tin.

Abundant mineral wealth in a region is often understood by scholars as a resource curse. This is a thesis which "suggests that natural resource abundance generates a series of economic and political distortions which ultimately undermine the contributions of extractive industry to development" (Bebbington, Hinajosa, Bebbington, Burneo and Warnaars 2008: 890). While the resource curse thesis has explanatory value, it is important to recognize that "resources do not do anything by themselves but through the social relations that make them significant" (Coronil 2011: 243). In many ways, contemporary mining is very similar to mining in the colonial era in that it relies on a local exploited labour force and can have negative social impacts for local populations who do not benefit from the profits that mining generates. Bury (2005) describes how the expansion of gold mining in the Cajamarca region of Peru (see Figure 5.5) has further concentrated land tenure as mining companies are able to purchase local land rapidly and the vast majority of profits generated are enjoyed by corporations rather than the local people. At the same time, however, mining has increased investment in health care, education and reforestation programmes that do benefit local people.

The impacts of mining on households and communities can often be quite unequal and uneven, making generalization difficult (Bury 2005). Furthermore, some mining operations lack adequate regulation or environmental impact assessment. In 2010, the world was gripped by the highly mediated plight of 33 Chilean miners successfully rescued after 69 days underground, but the issue emphasized the question of the safety of miners and whether the Latin American mining industry was doing all it could to protect the lives and health of miners.

Figure 5.5 *Yanacocha open cast gold mine, Cajamarca, Peru*

Source: Euyasik, Wikimedia Commons (Creative Commons Attribution-Share Alike 3.0 Unported, 2.5 Generic, 2.0 Generic and 1.0 Generic).

Not only do local people often fail to benefit substantially from mining operations, but it is clear that recent operations are having devastating environmental consequences. Mining operations have been blamed for contaminating surrounding waterways, sediments and soils with mercury, arsenic and cyanide; for depleting local water sources through cyanide leaching; for destroying topsoils, forests and wildlife habitats; for producing air pollution through smelting; for leaving inadequately disposed of tailings and other waste on the landscape; and for creating acid mine drainage (AMD), which occurs when tailings and waste are exposed to rainfall and which can contaminate waterways with toxic heavy metals (Liverman and Vilas 2006; Bebbington *et al.* 2008; Garibay, Boni, Panico, Urquijo and Klooster 2011; Zarsky and Stanley 2011: 30).

In La Oroya, Peru, where there is a smelter operation, the air pollution is so severe that children have to be evacuated from the city during the day (O'Shaughnessy 2007, cited in Bebbington *et al.* 2008). Swenson, Carter, Domec and Delgado (2011) have released some alarming findings on gold mining in Peru. Deforestation caused by gold mining now exceeds that created by settlement and Peru imports a staggering 142 tons of mercury a year, a key input in gold mining, which ends up in waterways, sediments and the atmosphere. Between 2003 and 2009, 6600 hectares of land (equivalent to 12,000 football fields) was transformed from forests and wetlands into mining wastelands. In that same period, the rate of deforestation increased by a factor of six.

Despite the creation of environmental codes of management, the environmental effects of contemporary mining operations appear to be far more severe than they were in previous decades and centuries. As Zarsky and Stanley (2011: 30) write with respect to gold mining in Guatemala:

> Rich veins of ore in which gold can be extracted in solid chunks have been exhausted. Today, gold is found primarily in low concentrations of less than ten grams per ton. To get the gold requires clearing vegetation and topsoil from large swaths of land; blasting large open-pit mines and underground tunnels and hauling the waste rock into large nearby mounds or valleys; excavating large amounts of ore and pulverizing it into a fine powder; treating the ore with a mix of water, lime and sodium cyanide; leaching the pregnant solution to separate the gold and sending it to a refining smelter, on or off site; and channeling the leftover tailings slurry to storage in a pond or "impoundment".

In other words, as Bebbington (2009) puts it, mining in Latin America has moved from open veins to open pits.

As a result of such negative environmental and social effects, political conflicts in mining areas are now extremely common throughout Latin America (Bebbington *et al.* 2008; Bebbington and Williams 2008; Urkidi and Walter 2011). For example, Urkidi and Walter (2011) discuss the movements for environmental justice that have emerged in both Chile (Pascua-Lama) and Argentina (Esquel) in opposition to gold mining. In Chile, a large Canadian company, Barrick Gold, is planning to start open-pit gold mining using cyanide leaching close to three mountain glaciers in 2013. Local people have mobilized against the Pascua-Lama project out of fears of water contamination and destruction of agricultural livelihoods. In the Esquel project in Argentina, a similar popular mobilization has emerged and the Mapuche people issued a declaration against US gold mining company Meridian, comparing this plunder of gold to that committed by the Spanish 500 years earlier (Urkidi and Walter 2011).

While the conflicts and protests surrounding mining are widespread, it is also important to recognize, as Bebbington *et al.* (2008) remind us, that mining provides employment and that Latin American mining communities, like mining communities all around the world, will often mobilize vigorously in defence of local mining operations.

Oil

Latin America is an important global producer and exporter of oil, a commodity so highly sought after that it never ceases to produce political and environmental conflicts. In the twentieth century, oil was discovered in a number of Latin American countries including Mexico, Venezuela, Brazil, Argentina, Peru and Ecuador. While oil provides a valuable source of export revenue, in order to extract it, Latin American countries have often had to allow multinational involvement, which has meant that many of the profits from oil are also exported.

As well as providing a precarious livelihood for the majority, the exploitation of oil has left a trail of environmental devastation. For example, Texaco began drilling for oil in the Ecuadorean Amazon in the 1970s, an activity that continued until the 1990s. For almost two decades, Ecuadorians have protested at the water contamination, toxic slime and high rates of cancer that the oil giant left behind. In 2011, a judge found Chevron (which purchased Texaco in 2001) guilty of widespread environmental damage to the Amazon basin and it was ordered to pay

$8 billion in damages (Rushe and Carroll 2011). In Colombia, oil exploration and extraction cannot be disentangled from the violence of the civil war and human rights abuses. Multinational oil companies are often targeted by both guerrillas and environmental activists. At the same time, those opposed to oil companies in Colombia have become the target of paramilitary activities and they have often been murdered by military protectors allegedly paid for by BP (Beder 2002; Pearce 2007).

Over the course of the past century, successive Latin American governments have attempted to gain leverage over their own oil industries by engaging in partial or full nationalization or increasing the tax rate when oil prices are high (Franko 2007). During the ISI period many Latin American governments nationalized their oil industries, while during the 1980s and 1990s under neoliberalism there was some partial privatization (Manzano and Monaldi 2008). Most Latin American governments have been reluctant to fully privatize their oil industries, and partnerships between multinationals and the state have been more common. Entities such as the Organization of Petroleum Exporting Countries (OPEC), of which Venezuela and Ecuador are members, have also enabled producer countries to exercise leverage over the international price of oil.

In more recent years, oil has been central to the resource nationalism being implemented by the left-wing governments of Bolivia, Brazil and Venezuela. Profits from oil are increasingly diverted into social programmes and national development initiatives. Venezuela's experience with oil revenue illustrates that it is possible to avoid the resource curse to some extent (Hammond 2011). Since 2007, vast quantities of new sources of oil have been discovered in Brazil, leading President Lula's chief of staff, Dilma Rouseff (currently the president of Brazil) to assert that it provided strong evidence that God is Brazilian (Duffy 2009). The current Brazilian government has publicly stated that it is committed to pouring oil wealth into social programmes.

Hazards and disasters

In addition to the environmental destruction brought about by colonial and neoliberal economic models, it is important to acknowledge that in geological and climatological terms Latin America is highly susceptible to a number of hazards. Half of the continent lies on a very active earthquake fault line, known as the Pacific Ring of Fire, and experiences hundreds of earthquakes every year. Some of these are devastating and cause massive loss of life, human injury and damage to buildings and property. In the past few decades, the continent has experienced a number of deadly earthquakes, including the 2010 Haiti quake, which killed 230,000 people and left half a million homeless.

In addition, much of Central America and the Caribbean is in a hurricane belt and usually experiences a number of tropical storms and hurricanes every year. Like earthquakes, some of these storms can cause catastrophic loss of life and other forms of devastation. The most devastating hurricane to hit Central America was Hurricane Mitch in 1998, which killed more than 11,000 people in Honduras and Nicaragua (see Box 5.4). In addition, the Pacific coast of the continent is prone to the El Niño Southern Oscillation (ENSO). Every five to seven years, a combination of higher than normal surface temperatures in the Pacific Ocean with high air surface pressure produces extreme weather, including droughts and floods, and can often impact negatively on farming and fishing.

Latin America also has a number of active volcanoes. In 1985, 25,000 people were killed by lahars created by the eruption of the Nevado del Ruiz volcano in Colombia. Melting glaciers are also serious hazards. People who live in Peru's Cordillera Blanca have repeatedly faced glacier disasters, such as glacial lake outburst floods, glacier avalanches or glacier landslides created by thinning or fracturing ice (Carey 2008). The 1970 Yungay earthquake triggered a glacier avalanche that buried tens of thousands of people. Box 5.3 is a selective list of the some of the major disasters to occur Latin America in the last few decades. It is important to recognize that in addition to the major events that attract (for a short period) international media attention, there are also many small-scale or slow onset hazards such as drought, desertification and air pollution that are often catastrophic in terms of their impacts on people and livelihoods.

Mainstream disaster management, planning and research have been dominated by a belief that disasters are created by extreme geophysical processes and can be solved by a technocratic approach and the implementation of engineering knowledge. Within this paradigm, little regard has been paid to the social construction of disaster and the way in which disasters are generated by economic and political factors and by unequal exposure to risk (Cannon 1994). Although the emphasis on external agents and natural forces continues to dominate disaster discourse (Hewitt 1995), disaster scholars are increasingly posing challenges to the dominant hazard paradigm because of its lack of social understanding. Cannon (1994) has suggested that the emphasis on the

Box 5.3

Selected list of large disasters in Latin America since 1960

2011 Brazil	Floods, 800 fatalities
2010 Haiti	Earthquake (7.0 magnitude), 230,000 fatalities
2010 Chile	Earthquake (8.8 magnitude), 525 fatalities
2010 Colombia	Flood, 418 fatalities
2009 El Salvador	Hurricane Ida, 275 fatalities
2007 Nicaragua	Hurricane Felix, 188 fatalities (see Figure 5.6)
2007 Peru	Earthquake (8.0 magnitude), 514 fatalities
2005 Central America	Hurricane Stan, 1600 fatalities
2001 El Salvador	Earthquake (7.7 magnitude), 1000 fatalities
1998 Central America	Hurricane Mitch, 11,000 fatalities (see Box 5.4)
1999 Venezuela	Floods (Vargas tragedy), 30,000 fatalities
1999 Mexico	Flood, 636 fatalities
1999 Colombia	Earthquake (6.8 magnitude), 1000 fatalities
1997–8 Andean region	El Niño, 600 fatalities
1997 Venezuela	Earthquake (6.9 magnitude), 81 fatalities
1997 Peru	Landslide, 300 fatalities
1994 Colombia	Earthquake (6.8 magnitude), 1000 fatalities
1992 Nicaragua	Earthquake (7.7 magnitude) and tsunami, 116 fatalities
1991–3 Whole continent	Cholera outbreak, 9000 fatalities
1991 Chile	Landslide, 141 fatalities
1988 Central America	Hurricane Joan, 200 fatalities
1988 Mexico	Hurricane Gilbert, 433 fatalities
1987 Ecuador	Earthquake (6.9 magnitude), 1000 fatalities
1987 Colombia	Landslide, 640 fatalities
1986 El Salvador	Earthquake (5.5 magnitude), 1000 fatalities
1985 Mexico	Earthquake (8.1 magnitude), 9500 fatalities
1985 Colombia	Volcanic eruption (Nevado del Ruiz), 25,000 fatalities
1982 Guatemala	Flood, 620 fatalities
1982 Mexico	Volcanic eruption, 100 fatalities
1976 Guatemala	Earthquake (7.5 magnitude), 23,000 fatalities
1974 Honduras	Hurricane Fifi, 6000 fatalities
1972 Managua, Nicaragua	Earthquake (6.2 magnitude) 10,000 fatalities
1970 Peru	Earthquake (7.9 magnitude) and avalanche, 70,000 fatalities
1961 Guatemala, Belize	Hurricane Hattie, 275 fatalities
1960 Chile	Earthquake (9.5 magnitude) 2000–6000 fatalities

Sources: US Geological Survey; EM-DAT; Guthmann 1995; Biles and Cobos 2004.

Figure 5.6 *A school in Krukira on Nicaragua's Atlantic Coast damaged by Hurricane Felix in 2007*

Source: Julie Cupples.

impact of nature can cause potentially dangerous interventions and argues that for a natural phenomenon to become a disaster, it has to affect vulnerable people.

In the words of Biles and Cobos (2004: 282), "[a]lthough damages are a function of the intensity and behavior of a physical and natural event, they are also directly correlated with the characteristics of territorial occupation". In other words, what is in place when the earthquake or hurricane occurs, including the levels of vulnerability and resilience of the affected population, will strongly influence the extent of the disaster that follows. So hazard exposure cannot be separated from vulnerabilities created by insecure employment and other elements.

It is also important to understand that people trade off different risks (Cupples 2004). People live by the riverbank in Matagalpa, Nicaragua in spite of exposure to flood risk as it is close to the market and provides informal employment opportunities. Some people who were relocated to safer ground and well-built homes by NGOs after Hurricane Mitch not only faced substantially longer travel times to the market, but also were at constant risk of muggings on their way home, which would result in the loss of any money or food they might be carrying. Some returned to live by the riverbank, feeling that the risk of losing daily earnings was greater than the risk posed by deadly but occasional flooding.

Disasters are, however, catalysts for political change. They mobilize survivors often in ways that could never have been anticipated, they reveal existing social, economic and racial inequalities and they often call the legitimacy of existing leaders into question (see essays in Buchenau and Johnson 2009; Solnit 2010). The 1972 Managua earthquake was a key contributing factor to the fall of the Somoza dictatorship in Nicaragua. The devastated survivors of the 1985 Mexico City earthquake, faced with government incompetence and corruption, began to engage in collective and creative acts of solidarity. This political mobilization would be the first in a series of factors leading to the gradual political decline of the governing party, the PRI, over the coming years, which culminated in its dramatic removal from office in 2000 after 70 years in power.

Box 5.4

Hurricane Mitch

One of the most tragic events to hit Central America at the end of the twentieth century was Hurricane Mitch. Ranked category 5 on the Saffir-Simpson scale, Mitch produced torrential rain for five days, washing away homes, livestock, roads and bridges and causing the deaths of at least 11,000 people. In Nicaragua, 3000 people were killed when a massive landslide swept down the sides of the Casitas Volcano, burying several communities that lay in its path (see Figure 5.7). Disease spread in the days after the hurricane owing to the contamination of the water supply by pesticides and chemical waste, by the decomposition of human and animal corpses and by the hundreds of latrines that had overflowed during flooding. Consequently, survivors also had to deal with widespread diarrhoea, respiratory infections, malaria, dengue, leptospirosis, cholera, conjunctivitis and tetanus.

Scholarly analysis of Mitch has been dominated by a political ecology framework because the hurricane clearly revealed both the extent of socioeconomic vulnerabilities and environmental degradation and how these processes intersected to turn a hazard into a major disaster (Bendaña 1999; Delaney and Schrader 2000; Cupples 2004, 2007b). A publication by the CCER (Civil Coalition for Emergency and Reconstruction), a coalition of 320 Nicaraguan NGOs created in the aftermath of Hurricane Mitch to coordinate and evaluate the reconstruction process, saw Mitch from that perspective.

Hurricane Mitch is the culmination of a long chain of impoverishment and deterioration of natural resources and of the quality and standard of living of the population. The devastating effects brought about by Mitch are closely related to the consequences of the historical model of development and the prevailing neoliberal economic model.

(CCER 1999: 18–19, author's translation)

Mitch's devastation cannot therefore be separated from what Mowforth (1999) refers to as the unsustainable nature of the operations of transnational companies in Central America. These involve the displacement of small farmers to the slopes of volcanoes and other marginal land; logging concessions that cause deforestation to advance at a rapid rate; mining operations that remove topsoils to extract minerals; and the diversion of rivers to irrigate huge banana plantations. All of these elements turned Mitch into one of the deadliest storms of the twentieth century and the magnitude of the disaster was compounded by other aspects of neoliberal economic policy, such as cuts to health care budgets and unemployment. In other words, making sense of Mitch requires an exploration of the linkages between neoliberalism, poverty and environmental degradation.

Figure 5.7 *A memorial park dedicated to the thousands of Nicaraguan victims of Hurricane Mitch is all that remains of the Community Rolando Rodríguez in Posoltega on the slopes of the Casitas Volcano*

Source: Julie Cupples.

Urban environmental issues

Urbanization rates are very high in Latin America and three-quarters of all Latin Americans live in cities. Latin America also has some of the world's largest cities. Consequently, the most pressing environmental issues facing Latin Americans are urban ones and have to do with access to clean water and sanitation, air and noise pollution and waste disposal. There are now more than 50 cities in Latin America with more than a million inhabitants, there are 10 cities with more than 3 million inhabitants and there are four mega-cities: Rio de Janeiro (12 million), Buenos Aires (13 million), São Paulo (22 million) and Mexico City (25 million) (Roberts and Thanos 2003).

The size of Mexico City, probably one of the most polluted cities in the world, makes its daily social reproduction unfathomable. The rapid growth of the city, industrial development, the poverty of many of its inhabitants and the wealth of others who were able to purchase cars created serious environmental problems. In addition to some 25 million inhabitants, Mexico City has some 3 million automobiles, buses and trucks which, along with industries, contribute to high and dangerous levels of air pollution and respiratory illness (see Figure 5.8). Respiratory illnesses are common in the informal settlements, some of which are located on dried-up lake beds and where the residents routinely breathe in dust and faecal matter.

Mexico City's geography makes air pollution more challenging to address. The city sits at an altitude of 2250 metres above sea level, which means the oxygen levels are lower, and it is also in a valley subject to an inversion layer that traps pollutants and prevents their dispersal. Mexico City is built on an underground aquifer which was an excellent source of water when the city was smaller. Today the aquifer is being rapidly depleted by a rate of extraction that exceeds replenishment so that the city is gradually sinking into the ground on which it was built. Water must be pumped up to Mexico City from lowland areas at great expense.

Latin American cities have grown very rapidly since the Second World War, although they have done so at different rates. While large first world cities such as New York and London took a long time to reach their current size, Latin American cities did so in the space of a few decades. Rural misery generated successive waves of rural–urban migration primarily in the second half of the twentieth century. Large numbers of Latin Americans settled in cities in search of employment, education and a better standard of living.

Figure 5.8 *The Zócalo in the heart of Mexico City, one of Latin America's megacities. Mexico City's inhabitants and workers routinely breathe in high levels of particulate matter*

Source: Marney Brosnan.

The rate of urbanization has been so rapid that urban authorities could not cope with the influx, so informal squatter settlements or shanty towns sprang up on the edges of all major cities. People built their own homes out of whatever materials they could access and gradually improved them over time. These informal settlements usually lacked access to water, electricity, sanitation, garbage collection, public transport, education and health care, which all became issues around which urban populations have mobilized and struggled.

Franko (2007) points out that one in three South Americans and one in four Central Americans live in city slums and are therefore exposed to substantial health risks associated with inadequate drinking water, poor sanitation, dumping of waste and substandard housing. Household fuel burning also poses serious health risks in the shanties, especially for women, as much cooking is done indoors using wood (see Figure 5.9). The inhalation of wood smoke causes serious respiratory illnesses.

In Latin America's informal settlements, water supply is both infrequent and erratic. It is often contaminated with microbes and parasites putting users, especially children, at risk of diarrhoea, which in some cases is life-threatening. Residents are often forced to buy water at exorbitant prices from vendors who sell it from trucks. One inhabitant of a Managua

Figure 5.9 *Cooking with firewood in Nicaragua*

Source: Julie Cupples.

shanty told me how the tap in her home is left permanently open and a metal container is left in the sink. If the water comes in at 3 or 4am, which it does about once or twice a week, the rattling of the metal container wakes up the household who then proceed to fill every single container and receptacle they possess before the supply runs dry again. This water must then be rationed and made to last until the supply returns.

Climate change

Many of these hazards and environmental problems are long-standing but they are now intensified by the question of global climate change. Climate change is not a new phenomenon. The historical record demonstrates that global climate has been prone to dramatic fluctuations, and previous periods of global warming and cooling have led to large-scale landscape transformations in many parts of the world. It is possible that climate change was responsible for the decline of several pre-Columbian civilizations. Drought was one of the factors that historians now believe contributed to the collapse of the Mayan civilization, the Classic Maya Collapse in the eighth and ninth centuries (Gill 2000; Webster 2002). The current period of warming is however increasingly considered to be one of the most serious issues facing global humanity, and large numbers of development agencies have turned their attention to mitigating climate change in the nations in which they work. Many scientists believe that the global climate is warming as a result of carbon dioxide emissions in the atmosphere and these emissions are responsible for generating catastrophic effects that are beginning to appear around the globe and are likely to get worse in coming years.

While global climate change is as much a first world issue as a third world one, there is a fundamental inequality at work in that the majority of climate change emissions are produced in the industrialized countries but wreak most havoc in third world countries. In the first world, climate change is often posited as a future-oriented problem, something that will affect future generations if we fail to act, while in Latin America it is affecting people right now (Cupples 2012). Despite this, there is a fundamental inequality at work in the generation of carbon emissions. Latin Americans use a fraction of the energy resources used by US Americans. While the average Chilean or Venezuelan uses 20 per cent of the electricity used by a US American, Guatemalans and Nicaraguans only use 2.8 and 2.6 per cent respectively (Franko 2007: 541). As we noted with respect to the deforestation of the Amazon, climate change should be approached as a development problem rather than as an environmental one (see Cupples 2012) – a challenge that the Bolivian government of Evo Morales has tackled head on (see Box 5.5).

Many Latin Americans are experiencing climate change directly. Subsistence farmers are more exposed to both drought and flooding and frequently lose their harvests as a result. The cultivation of biofuels from plants such as sugar and soybeans is posited as one of the "solutions" to climate change that overcomes oil dependence. Unfortunately for Latin

Box 5.5

Inaugural speech given by Bolivian president Evo Morales, People's Conference on Climate Change, Cochabamba, Bolivia, 20 April 2010

Brothers and sisters of the world, before I start, I want you to help me to say "planet or death!" Greetings to the vice-presidents of countries participating in this inauguration ceremony, to national and local authorities, to social movements of the world participating in this first international conference of people fighting for life, fighting for equality, fighting for dignity and the unity of humanity. . . . Copenhagen wasn't a failure, it's a victory for the people and a failure of the powerful nations of the world. In December 2009, the developed countries tried to get an agreement and thanks to your struggle, leaders of social movements who went to Copenhagen, together with some presidents, we tried to get across the feeling of suffering of peoples in the world, to get their demands heard. Those that tried failed, and so that is why we are meeting here now, because the so-called developed countries did not meet their obligations to establish substantial reductions in greenhouse gas emissions at Copenhagen. . . . Instead of saving humanity, the governments of the developed countries are going to allow global temperatures to rise more than 4 degrees Celsius: that is unacceptable, and that's why we've called this meeting. . . . If temperature is allowed to rise by more than 2 degrees, as established by the Copenhagen Agreement, global food production will fall by 40 per cent, world hunger will rise, and there are already 1 billion hungry people in the world and between 20 and 30 per cent of species will become extinct. Polar and glacier melting will intensify, and many islands will disappear beneath the ocean. . . .

In two and a half centuries, the developed countries which account for only 20 per cent of the world's population have generated 76 per cent of greenhouse gases. . . . I would say that the main cause of the destruction of planet Earth is capitalism; as people who inhabit and respect Mother Earth, we have the right, the ethics and the moral basis to assert that the main enemy of Mother Earth is capitalism. . . . Capitalism commodifies everything; water, land, ancestral cultures, justice and ethics. Until we change the capitalist system, the measures we adopt will have a limited and precarious effect. . . .

We need to create a new system that re-establishes harmony between nature and human beings. There can only be balance with nature if there is equity among human beings; there can be no harmony on a planet where 1 per cent of the population controls 50 per cent of the income. . . . Brothers and sisters, if we defeat capitalism, the task will be to look after Mother Earth with lots of love, maintain understanding, organization and unity for life and humanity. . . .

Long live the peoples meeting to defend Mother Earth!
Long live the rights of Mother Earth!
Death to capitalism!

Translation by author. The full text of the speech (in Spanish) is available at: www.komiteinternazionalistak.org

America, turning foodstuffs into fuel exacerbates unequal land tenure patterns and increases the cost of food and therefore hunger and malnutrition. Goodman (2007) refers to biofuels as the biggest scam going, and writes:

> Supporters of biofuel agriculture (grain and chemical companies, Wall St. investors, politicians and most University researchers) avoid mentioning the cost of inputs, the fossil fuels, the environmental damage, the physical toll on animals and humans, and the growing problem of hunger that will accompany the switch from food to energy crop production.

It appears that the cultivation of biofuels in Latin America merely replicates the inequalities, hardships and environmental damage produced by earlier forms of commodity trade, leading to hunger, displacement, violence and repression of smallholders, deforestation, pesticide use, nitrate runoff and water scarcity, and does nothing to halt the growth of automobile usage (Fitz 2007; Davies 2010; Bird 2011). Biofuel production is often more profitable for large landowners than food production, as they receive subsidies from the Clean Development Mechanism (CDM), an initiative developed by the United Nations Framework Convention for Climate Change which aims to reduce global carbon emissions (Bird 2011).

South America is currently gripped by soybean fever. The biofuel-driven "soyification" of Paraguay, for example, is having serious social and environmental consequences. By 2005, soybeans accounted for 10 per cent of Paraguay's GDP. US multinational GM seed and agro-chemical companies such as Monsanto, Cargill and Archer Daniels Midland as well as some Brazilian companies are gaining massive profits from Paraguayan soybean production. Paraguayan farmers who grow subsistence crops are being displaced by foreign soy farmers and deforestation of the rainforest to make way for soy planting has accelerated. Violent conflict between locals and foreign farmers is common. Large soy producers hire security firms to violently repress protests against the expansion of soy. Insecticides and pesticides used in soy production are also being blamed for a rise in birth defects and miscarriages and an increase in respiratory illnesses, headaches, skin rashes, vomiting and diarrhoea (see Monahan 2005; Howard and Dangl 2007; Barrionuevo 2008; Tana 2010).

The struggle for sustainable development

Despite the immense environmental challenges facing Latin America, many Latin Americans are working towards more sustainable ways of interacting with their local environments. In large cities such as Mexico City and Curitiba, Brazil (see Box 5.6), local governments and populations are promoting a range of sustainable practices. In rural areas,

---

Box 5.6

Urban sustainability

Curitiba is known as Brazil's ecological capital and is a model of sustainable urban planning admired around the world. In the 1960s, Curitiba, like other Latin American cities, was faced with the social and environmental challenges created by population growth, squatter settlements, poverty, pollution and traffic congestion. Jaime Lerner, a mayor with a vision, set out to create a sustainable city, taking all the city's problems head on and approaching urban sustainability in a broad way, thinking not just about environmental questions but also about the standard of living of the poor majority. This meant the creation of clean and efficient public transport, of new forms of waste management, of parks and other areas for recreation and of accessible forms of education.

These initiatives have been a huge success. The integrated bus system is so efficient that it is quicker and easier to travel by bus than by car and the vast majority of inhabitants travel around the city in this way. Shopping streets are mostly pedestrianized and there are many kilometres of bike lanes connecting parks and recreational areas. The Garbage Purchase Programme, which exchanges bags of garbage for school supplies, bus tokens and food, has enabled areas prone to dumping to be cleaned up and low-income families to improve their standard of living. The city also has a number of free educational centres. In these so-called *Faróis de Saber* (Lighthouses of Knowledge), people can access the Internet and library and cultural resources (see Rabinovitch and Leitman 1996; Irazábal 2005).

While Mexico City faces more substantial challenges than Curitiba, it too is embracing urban sustainability and a number of initiatives have been implemented since the election of centre-left PRD mayor Marcelo Ebrard in 2006. Mexico City was one of the first capital cities to sign up to the 10:10 campaign and commit to cutting carbon emissions by 10 per cent in one year. In addition to the underground system built in the late 1960s, there is now a Metrobus travelling in its own lane along Avenida Insurgentes, which overtakes the cars and saves 35,000 tonnes of carbon dioxide a year. Every Sunday the Paseo de la Reforma and other roads are closed to vehicular traffic and fill up with cyclists, roller bladers and skateboarders. Four urban beaches have been created right in the heart of the city to allow the poor of Mexico City to enjoy what the wealthy normally enjoy in places such as Acapulco. The local government is also replacing Mexico City's taxi fleet, building sustainable houses powered by geothermal energy and developing methane-capture programmes (see Harris 2007; Connolly 1999; Tuckman 2007; Moseley 2010).

local people are growing seedlings, reforesting hillsides and practising small-scale agro-ecological practices. These include companion planting, composting and the use of fertilizing bean to avoid the use of chemical pesticides and fertilizers, and building terraces on hillsides to prevent soil erosion. Demand for sustainably produced and fair-traded goods in first world countries is also providing opportunities. Some farmers are able to produce shade-grown organic coffee which they are able to sell to the fair-trade market for a premium, enabling them to protect the local ecosystem and provide a decent standard of living for their families.

The "pink tide" has allowed for a reorientation of some Latin American economies towards more sustainable modes of production. Venezuela, Ecuador and Bolivia have been moving towards the partial renationalization of natural resources and the channelling of resource wealth into social programmes that benefit larger numbers of people. Ecuador's 2007 cabinet included two ministers with activist histories of interrogating dependency on resource extraction (Bebbington 2009). That same year, the Ministry of Energy and Mines proposed to leave the petroleum reserves in the Yasuní National Park untapped as a means to preserve the area's rich biodiversity and keep 410 million tons of CO_2 emissions out of the atmosphere. It sought international financial help to be compensated for not profiting from the petroleum reserves. After several years of negotiations, Ecuador signed an agreement with the United Nations Development Fund in 2010 that will bring in funds from Germany, Spain, France, Switzerland and Sweden (Amazon Watch 2010). The deal came a year after Ecuador passed a new constitution giving rights to nature which include rivers, lakes, soils, forests, plants, birds and animals. In other words, nature is endowed with the "right to exist, persist, maintain and regenerate its vital cycles, structure, functions and its processes in evolution" and mandates that the government take "precaution and restriction measures in all the activities that can lead to the extinction of species, the destruction of the ecosystems or the permanent alteration of the natural cycles" (Mychalejko 2008).

In 2011, the Bolivian government passed the Law of Mother Earth, giving nonhuman living beings equal status with humans and enshrining the idea that nature has the right not only to exist but also "to not be affected by mega-infrastructure and development projects that affect the balance of ecosystems" (Vidal 2011). The significance of such changes notwithstanding, Bebbington (2009) does not think that post-neoliberal economies will be post-extractive economies, but the politics of contestation around extraction is generating new geographies of Latin America that require further consideration and research.

Conclusions

Engaging with Latin America's environmental issues means delving into a highly interdisciplinary project. In his work on banana cultures which draws together insights from history, geography, biology, agroecology, anthropology, political economy and cultural studies, Soluri (2005: xi) highlights "the need for people to think and act in ways that acknowledge the dynamic relationships between production and consumption, between people and nonhuman forms of life, and between cultures and economies". Indeed, his text shows us a way to study environmental problems that takes scientific perspectives into account but does not divorce "resources" from the social, political and cultural contexts in which they are exhausted, cultivated or extracted or from the people whose lives are inextricably entangled with them.

Latin America's struggle for sustainable development is a complex one. There are still too many factors working against sustainability. But amid the desperation, there are many sustainable initiatives that do get off the ground and do make a difference. The challenge for the future is to connect these struggles and put the continent as a whole on a less destructive path.

Summary

- The colonizers viewed Latin America as a site of natural resource exploitation.

- Neoliberalism has exacerbated environmental degradation.

- Deforestation of tropical rainforest became a global environmental issue in the 1980s for a number of reasons.

- Latin America is currently experiencing a mining boom, which is having negative environmental effects in many places.

- Oil extraction in Latin America has both increased and decreased poverty.

- Latin America has experienced many major disasters such as hurricanes and earthquakes.

- Hazard susceptibility is exacerbated by the neoliberal development model in place.

- Latin America's most pressing environmental issues are urban ones.
- There are many diverse initiatives under way to promote sustainable urban and rural development.

Discussion questions

1. What effect did colonialism have on Latin America's natural environments?

2. What is the relationship between neoliberalism and environmental degradation?

3. Why is it problematic to think of Hurricane Mitch as a *natural* disaster?

4. What are the factors working for and against sustainable development in Latin America?

5. What are the advantages and disadvantages of mining for local communities?

6. Why should climate change be considered a development issue?

Further reading

Carey, M. (2010) *In the Shadow of Melting Glaciers: Climate Change and Andean Society.* Oxford: Oxford University Press. This book provides a historical account of Peru's Cordillera Blanca and the ways in which a range of actors have engaged with the question of climate change.

Roberts, J. T. and Thanos, N. D. (2003) *Trouble in Paradise: Globalization and Environmental Crises in Latin America.* London: Routledge. A valuable introduction to a range of environmental problems in Latin America, including NAFTA, the Amazon rainforest and urban pollution.

Sawyer, S. (2004) *Crude Chronicles: Indigenous Politics, Multinational Oil, and Neoliberalism in Ecuador.* Durham, NC: Duke University Press. This book charts the destructive environmental consequences of multinational oil extraction and neoliberalism in Ecuador and the mobilizations by indigenous peoples against this state of affairs.

Soluri, J. (2005) *Banana Cultures: Agriculture, Consumption and Environmental Change in Honduras and the United States.* Austin, TX: University of Texas Press. An interdisciplinary focus on banana production and consumption and its social, political and environmental consequences.

Films

Maquiapolis – City of Factories (2006), directed by Vicky Funari and Sergio de la
Torre (USA). This film investigates the environmental devastation created by the
maquiladora factories in Tijuana, Mexico and the political struggle by local
women to get the factories to clean up their toxic waste and implement better
worker protection.

Crude: The Real Price of Oil (2009), directed by Joe Berlinger (USA). A film that
looks at the contamination of the Ecuadorian Amazon by multinational oil
company Chevron and the lawsuit that followed. www.crudethemovie.com

From Arbenz to Zelaya: Chiquita in Latin America (2009), Democracy Now interview
(USA). www.democracynow.org/2009/7/21/from_arbenz_to_zelaya_
chiquita_in

Bananas (2010), directed by Fredrik Gerttner (USA). A documentary about a group
of Nicaraguan banana plantation workers who sue the giant Dole Food Company
for using a banned pesticide known to cause cancer and sterility.
www.bananasthemovie.com

Bolivia: Fighting the Climate Wars (2011). A report by journalist John Vidal that
explores how climate change is impacting on ordinary Bolivians, resulting in
mudslides, agricultural failure, desertification and outmigration. It also shows how
the Bolivian government is speaking out on behalf of all poor nations affected by
climate change and denouncing the inaction of the first world on climate change.
www.guardian.co.uk/global-development/video/2011/apr/10/bolivia-fighting-the-
climate-wars

Radio programmes

Costing the Earth: Gold of the Conquistadors. BBC radio documentary that explores
the environmental impacts of the current mining boom in Latin America. First
broadcast BBC Radio 4, 12 October 2011. www.bbc.co.uk

Websites

www.emdat.be
The International Disaster Database. A collaboration between the World Health
Organization and the Belgian government, the database continues data on the
occurrence and effects of over 18,000 mass disasters in the world from 1900 to
present.

http://earthtrends.wri.org
World Resources Institute Earth Trends. A wealth of environmental information on
Latin America, including profiles, environmental impact assessment, data and
published reports, produced by the World Resources Institute.

www.conflictosmineros.net
Observatory of Mining Conflicts in Latin America (OCMAL).

6 Identity politics
"Race", gender and sexuality

Learning outcomes

By the end of this chapter, the reader should:

- **Be able to apply contemporary theoretical approaches to gender, race and sexuality to their manifestations in Latin America**
- **Be able to evaluate the role of identity politics in New Social Movements and development processes more broadly**
- **Be able to describe the progress made towards gender, racial and sexual equality in the continent**

Introduction

The colonization of Latin America was a gendered and racialized phenomenon. Very few Spanish women emigrated to Latin America and consequently, there were many coerced and consensual sexual relationships between Spanish men and indigenous women. These relationships produced large numbers of mestizo offspring whose fathers were absent and indifferent to their existence. The concept of the female-headed household, often understood as a more recent development phenomenon, has deep historical roots (Dore 1997). In addition, colonialism was an inherently racist phenomenon, based on a deeply held conviction that the colonizers were superior to the natives and the slaves, and were therefore justified in taking their land, their resources and in many cases their lives in order to fuel the development of Spain and Portugal.

The struggles for independence and the construction of new postcolonial nations were also embedded in dynamic and emerging forms of gender and racial subordination. Elites fighting for national liberation and independence were not interested in gender and racial equality. Indeed, while liberals and conservatives disagreed about trade and the role of the Church, they were united in their exclusion of women, slaves and Indians from formal political decision making (Appelbaum, Macpherson and Rosemblatt 2003). Nonetheless, both indigenous peoples and slaves resisted racism and discrimination in many ways. Brazilian slaves continued to practise their African religions and they developed a martial art form known as capoeira which could be disguised as a dance to trick the colonizers (Gates 2011). They also established their own fugitive communities which came to be known as *quilombos*. Gay and lesbian relationships were not endorsed by colonizers or the elites that ruled Latin America after independence and until very recently were usually kept hidden. Latin America has therefore come into being with very powerful yet intensely contested understandings of gender, race and sexuality.

Studying gender, race and sexuality has become more intellectually challenging in recent years. There is a large and growing body of literature on these questions in Latin America with different theoretical emphases and informed by different variants of feminist, anti-racist and queer theory. In an attempt to undermine essentialism, contemporary scholars have emphasized how gender, race and sexuality are socially and culturally constructed. Essentialism embodies the idea that gender, race and sexuality are biological and natural categories, a fact that justifies unequal status and treatment. On the other hand, a constructivist approach recognizes that these identity categories are not fixed or pre-given but rather they are brought into being through discourse and discursive practices. This means that they are both historically contingent and dynamic. They can be re-articulated by social actors to produce new modes of understanding, new outcomes and new forms of political intervention.

Today most natural and social scientists agree that race as a biological entity does not exist and there is no correlation between categories such as "black" and "white" and intellectual ability, and as a result many people prefer the term "ethnicity", which overlaps with race but is more about cultural location and belonging than phenotype (Wade 1997). But as Wade (1997) notes, race is still necessary as an analytical category because people behave as if races do exist and this behaviour often leads to discrimination and racism. Race is also evoked in less negative ways as an identity category around which people mobilize politically.

In addition to putting forward anti-essentialist understandings of identity, scholars have shown how class, gender, sexuality, race and ethnicity are mutually constituted. In simple terms, this means it is not the same to be a white woman as a black or indigenous woman and it is not the same to be a working-class woman as it is to be a middle-class woman. Any shared understandings that two women may have on the basis of gender are complicated by differential racial identities or class positions. Radcliffe and Westwood (1996) show how racialization, feminization and sexualization work to produce one another in the Latin American context in arenas such as domestic service. These socially constructed categories are both powerful and resilient, producing hierarchies and oppressions, but because they are socially constructed, they are always contested and unstable and can and frequently do change quite dramatically. Therefore, those that benefit from racial or gender inequalities are engaged in an ongoing ideological struggle to maintain such structures of privilege. At times, expressions of masculinity or femininity or racial identities change very rapidly and such change usually provokes widespread social anxieties in social groups that benefit from gender or racial inequality.

What we sometimes refer to as identity politics, or politics organized around particular axes of identity such as race, ethnicity, class or gender, has been central to development processes in Latin America. Social relations in Latin America from the colonial period until the present day have been profoundly shaped by hierarchies of race, class, gender and sexuality. These hierarchies and the ways in which individuals find themselves located within these arrangements affect subjectivities and how people construct and make sense of their identities and the development process. Their discursive power, along with their material manifestation in the form of racism, sexism and homophobia, means that they often constitute a barrier to participation and citizenship but that they can also serve as important vehicles for political mobilization. They have formed the basis of "new social movements" (Escobar and Alvarez 1992), which have provided new ways of doing politics and have promoted democratization in the region. New social movements (NSMs) refer to the ways in which Latin Americans began to organize politically outside formal or conventional political structures such as political parties or trade unions to form grassroots political movements based on gender, ethnicity, sexuality or neighbourhood. They include women's and feminist movements, indigenous movements, movements of *campesinos*, lesbian and gay movements and neighbourhood lobbying organizations.

NSMs, according to Asher (2009: 17), differ from conventional political movements because "they neither seek inclusion in existing political structures nor participation in leftist revolutionary strategies" and they seek instead to "resist the homogenizing forces of modernization and economic globalization and to imagine alternatives based on local knowledge". These organizations have tended to be more spontaneous, participatory and democratic than political parties and trade unions and have therefore appealed to wider sectors of the population.

In the 1970s and 1980s, important NSMs emerged largely in response to three main dynamics – the repression and human rights abuses enacted by dictatorships and military governments in Central and South America and around which people mobilized in defence of life, dignity and justice; rapid urbanization which led to informal and squatter settlements that lacked basic amenities; and the hardships and suffering created by neoliberal economic policies that led to a range of creative political responses. While class continues to be central to political struggles in Latin America, identifications relating to gender, race and ethnicity are now equally salient and in many ways have been disruptive to class as the central axis of identity around which mobilization occurs. In recent years, Latin America has seen vibrant LGTB (Lesbian, Gay, Transgender and Bisexual) struggles which are making important inroads into public discourse and social policy, although serious and ongoing prejudices remain.

This chapter explores the relationship between identity politics and Latin American development, focusing on race, gender and sexuality. While questions of indigeneity and indigenous struggles are pertinent to debates around both race and identity politics, their centrality in the contemporary Latin American conjuncture warrants a separate discussion in the next chapter.

Race

Nation building

After independence, the Latin American ruling classes were faced with the task of creating new nations and constructing coherent national identities. Latin American independence coincided with the emergence of nationalism and the notion of the nation-state in Europe. It also coincided with the emergence of scientific racism and social Darwinism, which understood race as a biological fact and asserted that white Europeans were innately more intelligent and more capable of civilization. The ideas of European racist thinkers such as Count Gobineau, Herbert Spencer, Gustave Le Bon and Georges Vacher de Lapouge were imported into Latin America and posed a dilemma for ruling elites. While these theories later became heavily discredited, they were very popular at the time of independence and as a consequence Latin American intellectuals such as Carlos Octavio Bunge, José Ingenieros, Alicides Arguedas and Euclides da Cunha began to express concerns about the continent's racial diversity. It was believed that the presence of large black and indigenous populations constituted an obstacle to economic development and progress. Scientific racism theorized that the whiter a nation was, the better it would be – an idea that took hold in many Latin American nations.

Liberal Argentine statesman Domingo Faustino Sarmiento was in many ways a progressive for his time, strongly promoting public education and democratization. At the same time, he believed that Latin America's social ills could be attributed to miscegenation with indigenous peoples, and modernization of the nation depended on promoting immigration from Europe (Stavenhagen 1994). Consequently, he also "called for the extermination of those who could not be educated, especially the indigenous peoples of the pampas" (Appelbaum et al. 2003: 5). Similar concerns were expressed by Simón Bolívar (Holt 2003).

The whitening, or *blanquemiento/brancamento*, of the population thus became an official policy in many Latin American countries in the nineteenth century. Brazil, Cuba and Argentina all set out to whiten their populations, but only Argentina succeeded in this endeavour and as a result it subsequently began to glorify the Europeanization of the Argentine nation. In the nineteenth century, Argentina began to encourage European immigration. Italians and Spaniards moved there in large numbers, but were joined by French, German, Russian, British and Irish immigrants. Unlike Mexico, for example, Argentina is frequently not understood to have a majority mestizo population.

Official whitening policies were accompanied by a discourse of racial democracy, to create an image of racial harmony and absence of racial tension (see Box 6.1). Racial democracy is an attempt to mobilize official discourse in such a way that racism is rendered irrelevant. The propagation of such narratives meant that race relations in Latin America took a different trajectory from those in the United States. Rather than mobilizing a black/white binary, in Latin America race relations were based on a colour continuum whereby the lighter you were, the greater your social status (Wade 1997, although see Skidmore 1993 and Lovell 1999, who call into question the idea of a bipolar US and multiracial Latin America).

Such understandings of race encouraged *mestizos* and mulattos to identify with the white elite and against Indians and blacks (Graham 1990). This approach did little to undermine the discrimination and marginalization experienced by black and indigenous populations and in turn it also did little to promote the loyalty of these populations to the nation. Racial democracy did however become a pervasive foundational

Box 6.1

Racial democracy in Brazil

Five million African slaves were shipped to Brazil during the colonial period and it took 350 years for the slave trade to be abolished. In the early part of the twentieth century, the Brazilian government pursued a policy of whitening to increase the percentage of whites, and some hoped that the blacks would eventually disappear. To accelerate the whitening process, European immigration was encouraged and mass sterilization campaigns targeted at Afro-Brazilian women were organized.

By the 1930s, it was apparent that such policies had failed and popular sentiment began to turn against European immigration. There was therefore a need to rethink Brazil's multiracial and multicultural condition, and so a number of Brazilian intellectuals and politicians began to assert that Brazil was a racial democracy. In other words, it was stated that through openness to mixing and intermarriage, Brazil had avoided the segregation and the racial tensions that are present in the United States and elsewhere. In the 1930s, Gilberto Freyre argued that racial divisions had become irrelevant and Brazil was becoming a culturally integrated nation (Rowe and Shelling 1991).

Since that period, racial categories have proved to be extremely fluid and difficult to pin down. It is difficult to assert who is and isn't black in Brazil. In popular discourse, Brazilians are known to use a large number of racial/skin colour categories to identify themselves or others, including cinnamon, toast, coffee with milk, and wheat (Grillo 1995). Many people with African ancestry do not identify as black in Brazil, and there is a close relationship between race and class. It is said in Brazil that "money whitens", so it is easier for wealthier people of African descent to identify as white or mixed race (*moreno* or *pardo*) (Travassos and Williams 2004). The idea of racial democracy became very powerful – so powerful that until recently it became difficult to acknowledge the existence of racism.

There is no doubt, however, that there is endemic racism in Brazil. Blacks are conspicuously absent from positions of power, including government and military positions. With many employers preferring to hire white employees, blacks earn less than whites and are much more likely to be unemployed or incarcerated. Infant mortality rates are higher among the black population. Blacks are also largely absent from Brazilian entertainment media, advertising and fashion catwalks.

In recent years, the myth of racial democracy has come under challenge from scholars and activists, including members of the Unified Black Movement (MNU). Black activists are finding new ways to express grievances and challenge racist practices and to "take control of the production of blackness" (de Santana Pinho 2010: 126). The 2010 census was the first time that a majority of Brazilians declared themselves to be black or mixed race, possibly suggesting a growing willingness to identify with African ancestry (Phillips 2011).

notion and led to a mainstream denial of the existence of racism. Because race was framed as a continuum rather than a binary, it became possible to transcend one's racial status, becoming more "white" by moving to the city, getting an education and adopting particular forms of dress, speech and occupation, a phenomenon known as "passing" (Appelbaum *et al.* 2003). It is a contradictory phenomenon, because on one hand it allows for some measure of social mobility, but on the other it reproduces the perceived superiority of whiteness.

Despite the dominance of this kind of racial thinking, opposing trends were also observed. The recognition in some places that whitening policies were doomed to failure opened the way for the accommodation, glorification and celebration of racial mixture or *mestizaje*. This process was particularly marked in Mexico after the Mexican Revolution (1910–1920), by which time most Mexicans self-identified as *mestizos*.

Mexican philosopher José Vasconcelos coined the term "cosmic race" to refer to Mexicans. This was scientific racism in reverse, which insisted instead on the superior nature of the hybrid. The mestizo thus became "a higher synthesis, neither Indian nor European, but quintessentially Mexican" (Knight 1990: 85). These ideas were similar to those being propagated by intellectuals elsewhere in the continent, such as Gilberto

Figure 6.1 *Afro-Mexican girl from Punta Maldonada, Guerrero, Mexico*

Source: Alejandro Linares García, Wikimedia Commons (Creative Commons Attribution-Share Alike 3.0 Unported).

Figure 6.2 *Afro-Mexicans protest their exclusion from the Mexican census*

Source: Israel Reyes Larrea, Wikimedia Commons (Creative Commons Attribution 3.0 Unported).

Freyre in Brazil and José Uriel García in Peru. In Mexico and Peru, this new nationalism was accompanied by a policy of *indigenismo* which aimed to integrate or assimilate indigenous Mexicans into mestizo national culture. These policies have had very contradictory outcomes for Mexicans in terms of national identity. Indeed, being Mexican means having "some Indian blood, but social aspirations require that they should not have too much" (Pitt-Rivers 1967: 547).

There were also important intellectual moves to revalorize black and African culture. In 1918, Afro-Brazilian scholar Manuel Querino published the first history of Brazil written from an Afro-Brazilian perspective, which documented the contribution made by blacks to Brazilian society, including Africans' skills in agriculture, mining and crafts. As Hamilton (2008) remarks, his work was provocative because he presented the African as a settler and as a collaborator with the Portuguese, but not as submissive in the face of enslavement. Despite the significance of Querino's work in reevaluating the black condition in Brazil, the white Freyre is far more cited and more remembered (Gates 2011).

In Cuba, a negritude movement developed similar to that elsewhere in the Caribbean. Negritude (also referred to as negrismo in Cuba) encompasses a conscious cultural engagement with Latin America's African heritages in literature, poetry, music or painting. In the 1920s, after centuries of colonial and postcolonial discrimination, Afro-Cuban culture began to flourish. Cuban poet Nicolás Guillén explored the cultures and everyday lives of Afro-Cubans in his work to develop new and more empowering ways of imagining and engaging with racial difference and conflicts in Cuba. Much of his poetry was written in the everyday vernacular of the Afro-Cuban population and emphasized *Cubanidad* (Cubanness) as a fusion of Spain and Africa. When *Motivos de son* were published in a Havana newspaper in 1930, the poems had a dramatic effect on the national psyche. As Augier and Bernstein (1951: 32) wrote:

> For days the pages of all Cuban newspapers were filled with the event. It was no trifling matter, we were in the presence of a phenomenon that established basic, irrefutable concepts in our poetry. Every voice in authority dealt with Guillén, either to applaud him or denounce him for his amazing feat.

Blackness came to be understood differently in different parts of Latin America. Even Haiti and the Dominican Republic, whose populations are mostly descended from African slaves and which are countries that share the same island, have developed quite divergent understandings of race. While the Dominican Republic has embraced its Spanish and Catholic heritage, leading many Dominicans not to see themselves as black, Haiti is proud of its black and African heritage. The voudou religion, feared and demonized elsewhere, is a source of pride, empowerment and community cohesion in Haiti. The longstanding historical tensions between Haiti and the Dominican Republic, which have frequently had tragic consequences, are mobilized through racial prejudice. Dominicans frequently project a racialized otherness onto Haitians. Indeed, the construction of Dominican national identity cannot be understood without reference to attitudes about race, Haiti and Haitians. In 1937, President Trujillo ordered his troops to massacre Haitians and later many Haitians became exploited sugar plantation labourers in the Dominican Republic. Anti-Haitian feeling is still strong in the Dominican Republic today (see Gates 2011 for discussion).

Race and racism in the twenty-first century

Racism, the privileging of whiteness and the myth of racial democracy continue to hinder progress towards racial equality in Latin America. Employers at major banks and hotel chains are much more likely to hire lighter skinned people as receptionists and other employees, while those who appear in advertising and entertainment media are generally much lighter skinned than the general population. It also results in racism expressed by *mestizos* towards blacks or indigenous peoples.

While many indigenous groups are mobilizing, in some cases to great effect, and these will be discussed in the next chapter, black Latin Americans or Afro-Latinos are gaining new forms of social and political visibility (see Box 6.2) in spite of ongoing forms of discrimination. African culture is highly visible in Brazilian society. As de Santana Pinho (2010) writes, these movements are also profoundly contradictory. In Bahia, often referred to as the most African part of Brazil, the inhabitants are reinventing blackness in creative and self-affirming ways and by so doing are attracting African American tourists to the region. The tension between the need to reject essentialist understandings of race and adopting strategically essentialist modes of valorizing blackness as a means to contest racism is, however, highly complex. De Santana Pinho argues that the reinscription of Africa on the body in a way that restores dignity also paradoxically tends to fix the black body and convert it into a site of commercial exploitation.

But in the twenty-first century, it appears that these discourses are now being more visibly and openly debated. In Mexico, the question of race was recently widely discussed after a controversy involving the United States erupted. In 2005, the Mexican postal service issued a series of postal stamps celebrating and commemorating a popular comic book character, Memín Pinguín. Memín Pinguín began in the 1940s and features the adventures of an Afro-Mexican child. It remains extremely popular with poor and working-class readers. The stamps provoked condemnation in the United States, with leaders such as Jesse Jackson, Al Sharpton and even President George Bush protesting that the images were racist. To many US Americans, Memín looked like Little Black Sambo and his mother like Aunt Jemima (Irwin 2009). Mexican leaders and Memín fans responded angrily, with some pointing out that the comic's content frequently contests racist behaviour. As Irwin (2009) notes, in one issue Memín is refused service in a Texan ice cream parlour because of his skin colour, so his friends cause serious damage to the store and are arrested for their actions.

At the same time it is important to recognize, as discussion of the debate acknowledges, that anti-racist discourse is often used by political leaders to mask and obscure racism (Pellow 2007; Irwin 2009). Indeed, the Memín postage stamps were released just a few weeks after Mexican president Vicente Fox had publicly stated that Mexican immigrants in the United States were only doing the work that "not even blacks" wanted to do, and some prominent and politically active Afro-Mexicans also believe the image is an offensive stereotype (Irwin 2009).

Box 6.2

Afro-Colombians

In recent years, struggles by black social movements in Colombia have taken on a new vitality and vibrancy. After Brazil, Colombia has the largest population of African-descended people in Latin America. Estimates of the percentage of black Colombians are contested, ranging from as low as 4 per cent of the population to as high as as 45 per cent (Wade 2002).

According to one estimate, almost 80 per cent of Afro-Colombians live below the poverty line (compared with a national rate of just under 50 per cent) and they are underrepresented in formal political spheres and university education (André 2011). Until recently, Afro-Colombians have also tended to be absent from official histories of Colombia. In the 1980s, however, attempts to reclaim blackness became part of the region's cultural politics. The 1990s provided further opportunities for empowerment and development. With the 1991 Constitution and Ley 70, Colombia "passed some of the most progressive legislation in the world for guaranteeing the collective property rights of its Afro-descendant minority population" (University of Texas 2007: 2). The constitution established Colombia as a multiethnic and pluricultural nation, while Ley 70 granted a number of social, economic, environmental and cultural rights to black communities, including the right to seek title to collective land ownership (Asher 2009; University of Texas 2007).

At the same time, however, the Colombian government was also pushing through large-scale development projects in agriculture, mining and tourism which have posed an obstacle to the securing of such rights. The expansion of oil palm in the region has been particularly conflictive. In addition, ongoing armed conflict waged by guerrillas and paramilitary groups in the region has forcibly displaced many Afro-Colombians from their homes and lands, and some black community leaders have been disappeared and murdered, further complicating attempts to seek legal title. Yet many Colombians continue to deny that racism exists in Colombia. Pacific Afro-Colombian hip hop group Choc Quib Town shocked Colombians with their public denunciation of Colombia as an institutionally racist country during their appearance at the 2011 Grammy awards in Las Vegas.

Persistent racism, poverty and marginalization have encouraged some Afro-Colombians to mobilize politically in defence of cultural identity, territoriality and biodiversity and against state-led developmentalism, although these struggles are not without their contradictions and compromised forms of political autonomy (see Asher 2009). They have however coalesced in important social movements such as the PCN (Process of Black Communities), who, according to Escobar (2008: 217), can be understood "in terms of the crafting of individual and collective identities in local contentious struggles". Amid huge political and ecological challenges, the PCN has developed a range of cultural initiatives that Escobar (2008) calls counterwork. These initiatives are creating significant challenges to the modernist, capitalist and destructive models of development led by the Colombian state. They include cooperatively organized coca and coconut production, a project to defend the biodiversity of the region and a popular communications project aimed at promoting oral cultures and alternative modes of literacy and expression and valorizing blackness.

While Memín is popular among subordinated populations and anti-racist readings are discursively available in the comics, there is no doubt that Afro-Mexicans remain politically invisible in Mexican society and Mexicans do not celebrate their African heritage in the ways they celebrate their indigenous ones. There are between 250,000 and 500,000 Afro-Mexicans, located mainly in the states of Oaxaca and Guerrero (see Figure 6.1). This group was however virtually excluded from Mexico's bicentennial anniversary in 2010 (Godoy 2009), in spite of the fact that many leaders of Mexican independence, including Vicente Guerrero, Juan Alvarez and José Maria Morelos, were descended from Africans (Foley 2010).

Afro-Mexicans have also recently protested their exclusion from the Mexican census. Figure 6.2 shows a demand to be counted as blacks, because as stated: "We are proud of our culture and of what we have given to this great Nation". Increasingly, though, one finds cultural events that both foreground and celebrate Afro-Latinos. In 2011, for example, Mexico City's National Museum of Popular Cultures in Coyoacán held a photographic exhibition dedicated to Afro-Mexicans, and in Venezuela, community radio is becoming a key site for the discussion of blackness (Fernandes 2010; see Chapter 8).

Recent migratory flows are also having an impact on the racial imaginaries at work in parts of the continent. In recent years, for example, the ethnic composition of Argentina has shifted quite substantially as a result of migration. After the collapse of the Soviet Union in the 1990s, Argentina experienced an influx of migrants from Eastern Europe. Today it is receiving many migrants from Peru, Bolivia and Paraguay that are more likely to identify as mestizo or indigenous than as white.

Gender

Latin American culture is rich in potent gender imagery and gendered iconographies. Gender, like race, has been central to national imaginaries in Latin America, which is known in stereotypical ways from outside the continent for its own particular brand of sexism, *machismo*. Latin America in fact mobilizes a diverse range of gendered icons, which draw on Catholic and colonial, pre-Columbian and supernatural heritages and revolutionary struggles. These include the Virgin of Guadalupe; Malinche, the mistress of Francisco Cortés and mother of the first Mexican; the weeping woman (la llorona), Catrina (Mexican death woman), the revolutionary guerrilla fighter, the caudillo and the hardy gaucho. These ideologies and iconographies, while powerful, are constantly subject to redefinition and rearticulation by writers, mediamakers and ordinary people in written texts, the media and in the spaces of everyday life.

Latin American gender relations are dominated by two pervasive gender ideologies – machismo or the cult of virility, which refers to males, and its female counterpart, marianismo, based on the cult of the Virgin Mary. Although the ideologies are meant to be complementary, they are based on a profound contradiction. While the machismo paradigm demands that men be sexually promiscuous and father as many children as possible, women are supposed to be sexually pure, chaste, submissive and self-sacrificing. Women are revered either as virgins or particularly as mothers. It is when women adopt the submissive position of the Virgin Mary that they achieve the greatest respect. But in the Latin American context it is accompanied by degradation, as it only by being submissive that they are idealized.

Women are often understood in dichotomous ways; the good woman or mother has her counterpart, the bad woman or whore. Of course, these cultural norms do not correspond neatly to the lived realities of individual men and women. Nevertheless, their power lies in their ability to inform and dominate people's subjectivities.

While Latin American women have endured subordination and oppressive gender relations in a range of ways, they also have a long history of political resistance and feminist struggle (Jaquette 2009). Throughout the colonial and postcolonial period, indigenous, slave and mestizo women organized to resist colonialism, patriarchal domination and the Catholic church.

Women, gender and development paradigms

When development first emerged as a discourse and an organizing paradigm in the second half of the twentieth century, development theory was largely gender-blind. The differential impacts of the development process on women and men were ignored, and if women were taken into account it was as mothers rather than as heterogeneous social actors in their own right. It was not until the 1970s that substantial shifts in thinking with respect to the relationship between women, gender and development could be observed. These shifts led to a number of paradigms, including Women in Development (WID), Women and Development (WAD) and Gender and Development (GAD), as outlined in Rathgeber (1990).

By the 1970s, particularly after the publication of the groundbreaking *Women's Role in Economic Development* (Boserup 1970), there was growing awareness that the development process was affecting men and women differently and that existing forms of gender inequality could be exacerbated by development interventions. WID, as Rathgeber (1990) notes, was a development variant of liberal feminism and a gender variant of modernization theory, which aimed to integrate women into economic systems, rather than attempt to understand why development had impacted negatively on women. WID had serious drawbacks, as it constructed women as victims, did not challenge the sources of women's subordination and failed to take account of the reproductive aspects of women's lives.

As the limitations of the WID approach became clear, WAD, a Marxist feminist paradigm, emerged in the second half of the 1970s. WAD intersected with dependency theory thinking and asserted that women had always been part of the development process but that women would not benefit until international systems were more equitable. While an improvement on WID, WAD also focused on women's productive activities at the expense of the reproductive side of women's lives and it also tended to construct women in homogeneous and essentializing ways as victims. These approaches sat uneasily with theoretical debates within feminist and gender studies which were focused on social constructionism and anti-essentialism and saw masculinity and femininity as relational constructs.

By the 1980s, GAD emerged. GAD had a much more holistic focus than either WID or WAD, focusing on both productive and reproductive activities, on relations between men and women, on the role of economic and political factors in shaping women's (and men's) lives, on gender as a socially constructed entity, and on women as agents of change rather than victims. GAD also advocated gender mainstreaming – the idea that gender should be central to all government and non-governmental institutions and institutional cultures. GAD overcame some of the problems identified with WID and WAD, but contained a number of limitations of its own (see Box 6.3).

In the latter half of the twentieth century, Latin American women mobilized in diverse and courageous ways: to overthrow dictators or contest human rights abuses, to confront economic hardship caused by economic globalization and to call for an end to domestic violence. They became part of local and transnational women's and feminist movements to fight for change on gender-specific issues such as reproductive rights and domestic violence. They also joined the labour market in large

Box 6.3

Making a development category: the female-headed household

One of the difficulties associated with GAD is that, like development more broadly, it creates its own problems for which solutions have to be found. In the late 1970s and 1980s, GAD research was pointing to growing numbers of female-headed households in Latin America and elsewhere in the developing world. Not only were their numbers growing, said the researchers, but households headed by women were also poorer and children growing up in such households were therefore at a disadvantage. Authors of such assertions thus constructed female heads of households as a vulnerable group that to date had been ignored in development planning but now required special targeting (Varley 1996). Researchers began to attribute the problem to economic modernization and rural–urban migration.

In 1991, the Chilean government created a government agency, SERNAM, that aimed to promote women's development. Taking on board GAD concerns, one of its initial priorities was the establishment of a programme for female heads of households, which would offer free training and childcare for women in this category to get them out of poverty. SERNAM discovered that it was difficult to find volunteers to take part in the scheme because the concept of female-headed households was largely unknown, and even women who were raising children without male partners did not identify themselves in this way (Valenzuela 1995).

Soon a body of more nuanced scholarship on female heads of households began to emerge. While this scholarship did not necessarily deny the growth in such households, it did point to their heterogeneity and diversity, in particular the diverse routes into household headship, the many different kinds of female-headed households and the fact that these households were not necessarily poorer than those headed by men (Varley 1996; Chant 1997a, 1997b; González de la Rocha 1999, Datta and McIlwaine 2000). Chant's (1997a) work for example has suggested that while men might earn more than women, they are also likely to indulge in what she refers to as extra-familial masculinities, such as drinking, gambling and extra-marital affairs. There is evidence that women are more likely to spend the money they earn on their children, so it cannot be universally concluded that these children are worse off (see also Cupples 2002, 2005).

It appears, as Peters (1995) has indicated, that when it comes to development practices, equating female-headed households with poverty and disadvantage has produced "analytical blinkers". Indeed, it has become apparent that underlying understandings of female heads of households as a vulnerable and problematic group is the assumption that the western ideal of the nuclear family based on a married heterosexual couple and their children is the best way to organize a household (see Harris 1981).

numbers, particularly as the New International Division of Labour (NIDL – see below) provided new sources of employment for women in free trade zones. This labour market participation also had significant implications for women's roles and identities. In recent years, the question of men and masculinities in development has become a salient issue with men's groups fighting against domestic violence.

Women in revolutionary movements

The revolutionary movements that swept Latin America in the twentieth century had substantial involvement of women. Women in Cuba, El Salvador, Nicaragua and Chiapas, many of them mothers, joined in militarized struggle to create a better world. While the actual numbers of women mobilized in revolutionary movements in Nicaragua and elsewhere are contested (Kampwirth 2002), there is no doubt that involvement in revolutionary struggles had significant implications for gender relations and identities (Randall 1981, 1994; Angel and Macintosh 1987; Collinson 1990; Kampwirth 2002; Cupples 2007a).

It is not the case however that military participation in revolutionary struggles automatically led to greater gender equality for women. The mobilization of women in revolutionary struggle led to an expectation that gender equality would be central to the political agenda of the revolutionary forces. However, in struggles for national liberation, gender-specific issues are often less prioritized than other issues. During the Allende government in Chile and the Nicaraguan Revolution in Nicaragua, gender issues such as domestic violence, reproductive rights and rape were dismissed as bourgeois, imperialist or culturally irrelevant by male revolutionary leaders (Molyneux 1986; Blandón 1994).

The Nicaraguan Revolution was particularly contradictory with respect to gender. In its 1969 programme, the FSLN stated that the abolition of sex discrimination was a fundamental aspect of the Sandinista struggle (FSLN 1969). After the triumph, the Sandinista discourse on women centred on the heroism of female commanders such as Leticia Herrera, Nora Astorga, Mónica Baltodano and Dora María Téllez (Field 1999), and gender equality was later enshrined in the 1987 Constitution. The Sandinistas also introduced a range of legislation that had the potential to vastly improve women's lives and created a mass women's organization, AMNLAE (Association of Nicaraguan Women Amanda Luisa Espinoza).

However, at the same time, the Nicaraguan Revolution continued to promote masculinist ways of doing development. Even AMNLAE's work was focused primarily on women's roles as mothers and tended to channel its activity towards associations of Mothers of Heroes and Martyrs which provided emotional and material support for women whose sons had been called up to do military service. Throughout the 1980s AMNLAE remained subordinate to the party and never became an autonomous feminist organization (Murguialday 1990; Lancaster 1992; Randall 1994; Blandón 1994). The reproductive powers of women were called upon by the FSLN to produce more supporters for the revolution, and motherhood and martyrdom, as many mothers lost sons in the Contra war, became central to revolutionary political activity (Puar 1996). According to Puar (1996), the FSLN publicly condemned machismo but by elevating motherhood, it was respecting its outcome. Many men who were good revolutionaries in the public sphere would be total reactionaries at home, controlling their partners' behaviour or indulging in domestic violence (Johnson 1985; Angel and Macintosh 1987; Lancaster 1992; Niehaus 1994; Randall 1994; Montenegro 1997).

In 1987 the FSLN issued a policy statement known as the *Proclama*, in which it acknowledged that very little had been achieved with relation to gender equality. It amounted to the first official recognition by the FSLN of the problem of domestic violence, as well as stressing the need for domestic chores to be a shared responsibility. However, it also acknowledged the role of Nicaraguan mothers as the "fundamental pillar of Nicaraguan family, defending it and sustaining it even in the most difficult circumstances, for which they deserve the highest respect and admiration" (FSLN Proclama 1987, in Puar 1996: 80).

Figure 6.3 *Comandante Ramona played a decisive role in the Zapatista uprising. She died in 2006*

Source: Heriberto Rodríguez, Wikimedia Commons (Creative Commons Attribution 2.0 Generic).

The Zapatista struggle in Chiapas appears to be somewhat different in terms of women's rights. The Zapatista rebellion is for Kampwirth (2002) "also a women's rebellion", not only because of the large numbers of women integrated into Zapatista forces (see Figure 6.3) but also because, unlike the revolutionary movements in Central America, it put feminist demands at the heart of its political agenda from the start. The Revolutionary Women's Law released in

1994 defends the right of women to make decisions in their communities, to control their own reproduction and to live free from violence (see also Kampwirth 2004).

Gender and human rights abuses

As we noted in Chapter 4, a number of Latin American nations were ruled by military governments and dictatorships during the 1970s and 1980s. The military's rise to power was generally accompanied by a call for a return to traditional values that it felt were being eroded, which included the idealization of "family values" and motherhood (Fisher 1993). But as Fisher (1989) writes, with respect to the situation in Argentina, while the military was extolling such values, beloved family members were being kidnapped, tortured and disappeared in a state-sponsored campaign of terror. Many women across the Americas, but especially in Argentina, Chile, El Salvador and Guatemala, lost husbands, sons, daughters and grandchildren to state terror and were forced to mobilize in response.

Schirmer (1988) has termed these human rights mobilizations motherist politics, as they were dominated by women searching for their missing children. These women frequently met in police stations, prisons, hospitals or morgues as they searched for any news of their loved ones. The most famous of these motherist groups are the Madres de la Plaza de Mayo of Argentina (see Figure 6.4), who have relentlessly protested the disappearance of their children every Thursday in the Plaza de Mayo in central Buenos Aires. Similar groups emerged in Chile (Group of Relatives of the Detained–Disappeared), Guatemala (Mutual Support Group or GAM; see Figure 7.1) and El Salvador (Co-Madres).

The military dictatorships were successful in closing down democratic political spaces, forcing women to find alternative channels of expression and protest. These governments aimed to depoliticize the populace and did not anticipate that this in itself would lead to new forms of collective struggle and opposition. What is particularly interesting about groups such as the Madres is that they were able to successfully turn the dictatorships' discourses on motherhood and the family against them. At a time when many people were being arrested and assassinated for involvement in political activities, these women presented themselves as good wives and mothers who were only trying to carry out their traditional roles in accordance with official ideologies. In effect, they agreed publicly with the government that motherhood was sacred and as a result the Catholic image of the suffering mother took on new and subversive meanings.

Figure 6.4 *The Madres de la Plaza de Mayo, Argentina*

Source: Julie Cupples.

They also took advantage of the traditional invisibility of women in the public sphere to oppose the dictatorships (Chuchryk 1989), an advantage that provided substantial if not complete protection from repression. The Co-Madres of El Salvador, for example, always dressed in black as a form of protection (Schirmer 1993), while the Madres of Argentina would pretend to be knitting or praying in church while they passed around messages (Fisher 1989). The belief that the military would not harm middle-aged women acting as mothers to defend their children empowered them to take action in dangerous political circumstances. As Noonan (1995) and Schirmer (1988) indicate, the discourse that aimed to oppress women and keep them in place proved instead to be an important source of female empowerment and political opportunity. Consequently, these women became politicized subjects without a necessary prior awareness of or opposition to gender inequality and without an intention to engage in feminist struggle (Fisher 1993; Guzman Bouvard 1994). In the process, the hypocrisy of the military governments with respect to family values was more than amply revealed (Feijoó and Gogna 1990).

Since these groups' creation, both the mothers and grandmothers of Argentina have fought to bring perpetrators to justice and to locate the remains of those disappeared. Their work has however diversified into other areas and the Madres have created a bookstore, a university, a library, a cultural centre, a newspaper, a radio station and a subsidized housing project. Their involvement in these political causes can be understood as an attempt to keep alive the revolutionary political aims that their children died for.

Neoliberalism and urban struggles

The harsh economic conditions generated by neoliberal economic policies had particularly gendered impacts. The philosophy behind IMF conditionality was that the cost of adjustment was to be borne by individuals rather than the state. Clearly, the increases in the costs of basic foods and services impacted most severely on the marginalized urban poor. Because women are generally responsible for reproductive activities, such as putting food on the table and caring for children and the elderly, women tended to be more severely impacted by these policies. Indeed, women became the shock absorbers of the crisis (Safa and Flora 1992) as they were forced to invent creative ways to reproduce the family. Many women joined the labour market for the first time, taking on jobs in the informal sector (see Figure 6.5), in domestic service and in assembly factories.

Figure 6.5 *The gendered informal sector: under neoliberal structural adjustment, many women went to work in the informal sector. A woman in León, Nicaragua sells raspados (snow cones) from a cart*

Source: Marney Brosnan.

In Chile, Peru and Uruguay, some women organized themselves into providing food for their families communally, creating community soup kitchens – the so-called *comedores populares* or *ollas comunes*. Others became involved in handicraft cooperatives or sewing workshops which provided new sources of income. A smaller number of women became active in squatter settlements or land invasions.

One of the interesting gender dimensions of activities such as communal kitchens is that as women found collective solutions to reproductive tasks that had become too onerous as a result of the economic situation, in the process they also began to address more strategic gender issues. For many women, participation in these community-oriented activities proved to be an important source of self-education and empowerment (Chuchryk 1989), dismantling any kind of binary separation between practical (providing food for families) and strategic (pursuing feminist aims) gender issues (see Molyneux 1986). Involvement in these activities led to a renegotiation of gender relations. Conflicts over household division of labour often resulted, and many women had to deal with opposition from husbands or other family members (Fisher 1989; Pires de Rio Caldeira 1990). For many men, having to get used to eating food cooked communally rather than by their own spouse was a profound cultural change (Sara-Lafosse 1989).

High male unemployment left many men with little option but to accept the situation. In the process, their role as main or sole income earner in the household was undermined and as a result, women began to question other aspects of male household authority (Safa and Flora 1992). These activities were highly contradictory. On one hand, women gained new political and organizational skills that were potentially transferable and new forms of bargaining power in their homes and communities. On a practical level, women could stop worrying about their children's nutritional intake and communal cooking meant that women did not have to cook every day, which freed up time for other productive or political activities (Barrig 1989; Sara-Lafosse 1989). On the other hand, communal cooking meant that food distribution remained the role of women, reproducing traditional gender roles. It also meant that women absorbed the shocks created by economic crisis, which partially relieved government of its responsibilities to its citizens.

Traditional constructions of masculinity and femininity in Latin America have equated men with the public sphere and the labour market and women with the private sphere and the home. In many parts of the continent, masculinity is equated with being a good provider, so if a woman does engage in paid work, the masculinity of her husband is called into question. Furthermore, women's paid work is often assumed to be supplementary to the male income. Rapid economic and political transformation in the continent has caused such ideologies to undergo quite substantial shifts. Of course, for many low-income families, women's paid work has long been essential to family survival and so the degree to which individuals subscribe to such ideologies is highly variable. The persistence of such cultural norms, despite widespread non-conformity, means that even though women are engaged in the labour market, they are often subject to forms of discrimination, earning less than men and often subject to less favourable working conditions.

The restructuring of the global economy and in particular the shifting of economic production away from national economies – the so-called New International Division of Labour (NIDL) – has had a decisive impact on women's labour market participation in Latin America (Stichter 1988; Standing 1989). The development in particular of assembly factories or maquiladoras (see Figure 3.5) along the Mexican side of the US–Mexican border and in free trade zones (FTZs) elsewhere in the continent has created new sources of low wage employment for young women in textiles, electronics and industrial agriculture. Through the creation of such feminized occupational niches, Latin American women have become tied into global production networks as the overall share of male employment has fallen. The preference for young, mostly single and childless women has been attributed to prevailing cultural understandings about gender that assume that young women are more docile and obedient, more suited to repetitive factory work and less likely to unionize, and that their incomes are supplementary to the household so they could be paid lower wages (Elson and Pearson 1981; Arizpe and Aranda 1986; Tiano 1990; McClenaghan 1997).

Early explorations of the relationship between gender and the NIDL tended to focus on exploitation, seeing women primarily as victims of industrialization (Benería, Floro, Grown and MacDonald 2000). While wages are generally low and working conditions often extremely poor, much of the early work, informed by political economy approaches, tended to construct women as victims of neoliberalism and globalization. This is how the late Anita Roddick described conditions at the factories in Las Mercedes Free Trade Zone in Managua:

> On Managua's outskirts is Las Mercedes Zona Franca, the city's free trade zone. Its factories employ some 22,000 people, most of them young mothers, who work ten hours a day, six days a week from 7.00am. Earning US$4–5 a day, they hope to take home US$130 a month after forced overtime – $70 less than the basic monthly wage which independent economic studies say is needed to survive. But with the country's unemployment rate a staggering 60 per cent, no viable alternative exists for many workers. The vast clothing factories are owned by Taiwanese, Korean and US firms, and manufacture jeans, shorts, trainers and shirts destined for sale in retail outlets in the US and Canada. The price mark-up is huge: for each pair of jeans sold in North America for $20–30, the workers receive around 20 cents. Workers complain of being physically and verbally abused by their bosses, likening their employment to a form of slavery.
>
> (Roddick 2001: 32)

While such criticisms are by no means inaccurate and scholarship critical of exploitation is essential, another body of gender and development scholarship informed by poststructuralism has instead identified the complex motivations, diverse experiences and forms of female agency that underpin women's factory work (see for example Laurie, Dwyer, Holloway and Smith 1999; Tiano and Ladino 1999; Benería *et al.* 2000), undermining the concept of a stereotypical third world factory worker. Such work has revealed that women's increased access to earnings has empowering implications at the micro-scale, enabling them to negotiate relationships more decisively, leave abusive relationships or replace unpaid domestic work with low-paid factory work.

Tiano and Ladino's (1999) work in Ciudad Juárez in the Mexican border region suggests that women do not experience negative impacts directly but rather these are mediated through gender identities that shift over time. Through their factory labour, women are exposed to alternative ideologies that challenge traditional understandings of Mexican womanhood. They reported being able to socialize more freely and choose mates without parental supervision. Even in the context of arduous work and low pay, their autonomy expanded, they gained access to greater sources of personal enjoyment than their mothers' generation and as a result their understandings of femininity, motherhood and domesticity were reconstructed in positive ways. Still, as women workers and cheap labour, women continue to be devalued, an attitude that has led to the femicide in Ciudad Juárez (see Box 6.4).

Export manufacturing has had mixed outcomes for women. There is a need to move away from analyses that try to decide whether export manufacturing is good or bad, and consider instead how these processes are mediated by gender identities. It is clear that constructing a low-waged female labour force is a complicated and contradictory process. Multinational companies profit from prevailing gender norms which make women disproportionately responsible for home and family and which restrict their access to alternative employment, but the apparent ability of the multinationals to take advantage of cheap labour for profit also brings its own contradictory outcomes.

Drawing on work conducted in the Dominican Republic, Raynolds (1998: 161) argues that multinational firms "engage in ideological juggling: balancing their efforts to recast women as acceptable wage workers with bolstering the gender subordination that makes women's labor inexpensive". It is therefore hardly surprising that women are able to find spaces of opportunity and empowerment amid the exploitation.

Box 6.4

The city of dead girls

Since 1993, hundreds of maquiladora workers have been murdered in Ciudad Juárez, Mexico, an export processing zone on the US–Mexican border. Their sexually violated bodies are usually dumped in the desert on the outskirts of the city. The perpetrators of such crimes are rarely brought to justice.

Garwood (2002) asserts that there is a close relationship between cultural, structural and physical violence at work in this situation. The women are subject to structural violence, through their exposure to poor working conditions and inadequate access to housing, health care, transport and potable water. They are also subject to cultural violence given the assumptions made by multinational firms about their worth and easy expendability. Their gender subordination as cheap female labour contributes to the physical violence – sexual violence and murder – that is an everyday reality in the city. This gender subordination also results in inadequate police investigation of those responsible along with suggestions that the women were responsible for their own deaths through immoral or inappropriate behaviour.

The maquiladoras implement quite intense gendered and sexualized forms of surveillance at work which extend to forced pregnancy tests, compulsory birth control and the organization of company beauty pageants, yet they are unwilling to provide adequate protection to women to get to and from work safely (Garwood 2002). While these women are making decisive contributions to the global economy, their contributions are invisible and instead they are discursively constructed as "cheap labour" and "loose women". The mutual constitution of these discursive constructions is what keeps their labour cheap and undervalued and allows for sexual violence to take place with impunity. As a result of the murders and police failures, a dynamic women's movement fighting for justice has emerged in Ciudad Juárez (see Figure 6.6).

Progress towards gender equality

Latin American women's and feminist movements have put gender-specific issues on national political agendas and by so doing have made real differences in women's lives. All Latin American countries now have legislation to deal with the issue of domestic violence, which has been an endemic social problem in many countries. For a long time, domestic violence was seen as a burden to be tolerated privately and in silence, but women's organizing and awareness-raising have brought the issue out into the open. Both victims of and witnesses to domestic violence are now much more likely to denounce such crimes than they were a few years ago.

More and more Latin American women are running for and being elected to office. Latin America has already had several female presidents,

Figure 6.6 *Cludad Juárez, Mexico: pink crosses mark places where young maquiladora workers were raped and murdered*

Source: Wikimedia Commons (released into public domain).

including Violeta Barrios de Chamorro of Nicaragua, Cristina Fernández de Kirchner of Argentina, Michelle Bachelet of Chile, Laura Chinchilla of Costa Rica and Dilma Rouseff of Brazil. Other gender issues are more difficult to address. In particular, little progress has been made on abortion rights. Abortion is still illegal almost everywhere in Latin America and consequently, women desperate to end unwanted pregnancies are forced to take grave risks with clandestine abortion providers.

Men and masculinities in development

The move from WID to GAD in the field of gender and development practice posited a move from women to gender. It constituted an attempt to disrupt the understandable yet overwhelming preoccupation with women and to encourage contributions from men. Yet most GAD texts continued to be about women and women's issues. While the literature and practice on gender and development inevitably referred to men or had implications for men and men's lives, men as men were missing from the picture (Cornwall 2000). As Cornwall (2000) writes, GAD tends to construct men and women in dichotomous terms, whereby women are hardworking, reliable, trustworthy, socially responsible, caring and cooperative, while men are lazy, violent, alcoholic and unfaithful. Men appear as obstacles to women's development rather than people with whom women might have shared interests.

There is however a growing body of scholarship that attempts to dismantle stereotypical representations of men as well as forms of development practices that are attempting to deconstruct violent or negative forms of masculinity. Gutmann's (1996) ethnographic study of a low-income neighbourhood in Mexico City reveals that in contrast to the "typical Mexican man" as a "hard-drinking philandering macho", men hold their children, play with their children, have a particularly important role in raising sons and view fatherhood as a lifetime commitment. Although there are normative differences in parenting responsibilities between men and women and clearly many men are complicit in sustaining hegemonic masculinity, non-conformity is always an option (see Connell 1995). Gutmann (1996) argues that undifferentiated and essentializing concepts of motherhood and fatherhood are unfounded and misleading. In fact, Gutmann found that the extent to which individual men subscribe to specific masculine attributes is highly variable and arbitrary. Some alcoholic men are good providers; some men abstain from drinking but are violent towards their wives; and some women beat their children or indulge in infidelity or drinking.

The image of the macho man is familiar but it is not fixed, and alternative subject positions are also discursively available to individual men (see Figure 6.7). In Nicaragua, the Group of Men against Violence, a nationwide group that seeks to tackle the problem of domestic violence, has recruited large numbers of men, many of whom have been violent towards their partners but want both to change personally and to bring out transformations in their communities (see Montoya Tellería 1998; Walsh 2001). These groups use popular education methodologies to explore and deconstruct expressions of masculinity that have violent consequences and are negative for their partners and children.

Sexuality and LGBT politics

While social movements for gender and racial equality have gained substantially in visibility and effectiveness over the past three decades, movements to secure the rights of Latin America's lesbian, gay, bisexual and transgender (LGBT) populations are more incipient. As Corrales and

Figure 6.7 *Masculinities in development: a Cuban father in Trinidad enjoys some time with his daughter*

Source: Marney Brosnan.

Pecheny (2010) write, while some important cultural and legislative advances have been made in some parts of the continent, these achievements are highly uneven.

Most Latin American societies have culturally entrenched forms of homophobia. Neither right-wing military dictatorships nor left-wing revolutionaries were LGBT friendly. A gay organization, the Frente de Liberación Homosexual (Homosexual Liberation Front) in Argentina was repressed by the military dictatorship that came to power in 1976 along with the other political groups that the regime deemed subversive. In the 1960s, the Cuban government even went so far as to incarcerate gay men in labour camps.

Today, fears of social and familial rejection work to keep LGTB people in the closet and deter them from political mobilization against heteronormativity and in defence of citizenship rights. While lesbian women have been largely invisible, gay men, who are deemed to have rejected dominant cultural expressions of masculinity, are often subjects of hostility, ridicule and even violence. Transgendered Latin Americans – people whose gender identity does not correspond to the gender assigned at birth – are similarly faced with substantial social misunderstanding.

It is important to note, however, that homosexuality takes culturally specific forms throughout the continent. Lancaster's (1992) ethnography of 1980s Nicaragua describes a particularly Nicaraguan form of stigma attached to homosexual practices. While it is important not to simplify or binarize homosexual desire or practice (see Palaversich 1999), according to Lancaster, only the passive (penetrated) male or *cochón* is stigmatized, while the active (penetrating) man is not, and indeed the latter can gain status by sleeping with many *cochones* just as one does by sleeping with many women.

Since the 1980s, Nicaragua has however changed dramatically and today there is substantial evidence of an increasingly visible, assertive, diverse and transnational LGBT movement and far greater willingness by Nicaraguans to engage in debates regarding (homo)sexuality (Babb 2009). This is something that can be observed across the continent. Important changes in attitude are accompanied by legislative and policy changes (see Box 6.5).

LGBT activists have made important inroads in deeply homophobic legislatures by framing their search for equality as a human rights issue and successfully connecting it to other human rights related struggles (Encarnación 2011). The fact that the struggle for human rights in countries such as Argentina has been so powerfully fought and felt, and so costly to so many people, probably makes more people receptive to gay marriage on the ground of human rights. It is clear that while legislative change comes in the wake of cultural change and political demands, legislative changes also foment cultural change and can make important contributions to social tolerance. A survey conducted in Mexico after the first same-sex marriages were celebrated showed people becoming much more accepting, with the numbers of those opposed to such marriages falling substantially (Vinter 2010). Increasingly, Latin American cities such as Buenos Aires and Mexico City emphasize their gay-friendliness to tourists and visitors.

There is still a long way to go in Latin America in overcoming homophobia, and some countries have rejected legislation passed elsewhere. Honduras and the Dominican Republic for example passed laws banning same-sex marriages and adoptions in 2005 and 2009 respectively, and violence and hate crimes against LGTB people are still horrifically frequent. Recent cultural and legislative changes are however impressive and provide an excellent basis from which to fight for full citizenship rights for LGBT citizens.

Box 6.5

Coming out in Latin America

1998, Ecuador	New constitution includes protections against discrimination based on sexual orientation
1999, Chile	Decriminalization of same-sex intercourse
2000, Brazil	The state legislature of Rio de Janeiro bans discrimination based on sexual orientation in public and private establishments
2003, Mexico	Federal anti-discrimination (including sexual orientation) law passed
2004, Brazil	Government introduces *Brasil sem Homofobia*, a programme to change social attitudes towards gays and lesbians
2004, Peru	Repeal of a law that banned members of the armed forces from having homosexual relations
2006, Mexico	Government passes a cohabitation law, which grants cohabiting same-sex couples the same rights as cohabiting heterosexual couples
2006 and 2007, Mexico	Coahuila and Mexico City legally recognize same-sex civil unions
2007, Argentina	Buenos Aires allows same-sex unions
2007, Brazil	Rio Grande do Sul allows same-sex unions
2007, Uruguay	All cohabiting couples given access to health benefits, inheritance, parenting rights and pension rights, after five years of cohabitation regardless of sexual orientation
2007, Colombia	Cohabiting same-sex couples given same rights as married couples after two years of cohabitation
2008, Mexico	Trans (transgender, transsexual, transvestite and other non-cisgender gender identities) persons permitted to change their legal gender and name on official documents
2008, Nicaragua	Decriminalization of same-sex relationships and repeal of anti-sodomy law
2008, Cuba	Government introduces free sex change operations
2008, Brazil	Three million people attend the LGBT parade in São Paulo, the largest gay pride march ever held (see Figure 6.8)
2008, Brazil	The First National Conference of Gays, Lesbians, Bisexuals, Transvestites and Transsexuals is held in Brasilia
2008, Panama	The government repeals a law that criminalizes same-sex intercourse
2009, Bolivia	New constitution bans discrimination on the basis of sexual orientation
2009, Chile	Social activists organize a mass wedding for sexual minorities in front of the Metropolitan Cathedral
2009, Colombia	Same-sex couples granted same pension and property rights as heterosexual couples
2009 Mexico	Legislature of Mexico City approves marriage and adoption rights for same-sex couples
2009, Uruguay	Government approves a bill that ends the restriction of adoption to married couples
2010, Argentina	Legalization of same-sex marriage – the first Latin American country and eighth in the world to do so

2011, Brazil	Supreme Court grants equal legal rights to same-sex civil unions as those enjoyed by married heterosexuals, including retirement benefits, joint tax declarations, inheritance rights and child adoption

Sources: *The Economist* 2007; Corrales 2010; Corrales and Pecheny 2010; Soltis 2011a.

Figure 6.8 *Gay pride parade in São Paulo, Brazil*

Source: Agência Brasil, Wikimedia Commons (Creative Commons License Attribution 3.0 Brazil).

Conclusions

Racism, sexism and homophobia are all still alive and well in Latin America. These oppressive forces should be understood as a development issue because they prevent people from reaching their full potential by denying them access to opportunities that are available to people who are differently racialized or gendered or who express a different sexual orientation. The struggles to change this state of affairs have however been truly inspiring. Latin America has a vibrant and sophisticated feminist movement that is making its presence felt in scholarship, in policy-making arenas, in the street and in the home. Latin America's black and indigenous (see Chapter 7) populations are also mobilizing and asserting their blackness and indigeneity rather than attempting to transcend it by "passing". Progress against homophobia is less well developed but the concept of human rights is one that many Latin Americans support, and it is likely that the progress made to date in LGTB rights will be built upon in the future.

As socially constructed identity categories, "race", gender and sexuality are both resilient and malleable, depending on the context in which they are constructed and contested. It is therefore impossible to generalize about Latin America as a whole and it is necessary to explore these categories in geographically specific ways, as many Latin Americanists have done.

Summary

- Colonialism was a gendered and racialized phenomenon.

- Identity politics and new social movements organized around race and gender have been central to development processes.

- Nation building involved the construction of quite specific ideas about race.

- There have been a variety of political and intellectual attempts to revalorize indigeneity and blackness.

- Latin American gender relations are dominated by machismo and marianismo.

- Women have participated fully in revolutionary movements but these movements have had contradictory outcomes for gender relations.

- In the 1970s and 1980s, many women organized as mothers to protest against human rights abuses.

- Women have organized in diverse ways in the face of economic crisis.

- Not all men subscribe to hegemonic models of masculinity.

- LGBT movements are less visible than other forms of identity politics but have made significant cultural and political advances in recent years. Homophobia is however deeply entrenched.

Discussion questions

1. Discuss the diverse ways in which race and racial identity have been understood in Latin America and how these understandings have changed over time.

2. What are the cultural and political advantages and disadvantages of mobilizing around black identities?

3. Study the three main development paradigms dealing with women, gender and development. How do they help us to theorize the relationship between gender and development in Latin America?

4. Discuss the relationship between women's involvements in revolutionary movements and the struggles for gender equality.

5. What are the contributing factors to the exploitative treatment and brutal violence experienced by maquiladora workers in Mexico?

6. Discuss how earlier Latin American struggles for human rights are shaping more recent struggles around sexual rights.

Further reading

Asher, K. (2009) *Black and Green: Afro-Colombians, Development, and Nature in the Pacific*. Durham, NC: Duke University Press. Drawing on ethnographic fieldwork with Afro-Colombians, this text looks at the relationship between development and social movements.

Chant, S. and Craske, N. (2003) *Gender in Latin America*. London: Latin America Bureau. A review of contemporary gender issues in Latin America, with chapters on politics, poverty, population, health, sexuality, families and households, employment and migration.

Corrales, J. and Pecheny, M. (eds) (2010) *The Politics of Sexuality in Latin America: A Reader on Lesbian, Gay, Bisexual, and Transgender Rights*. Pittsburgh, PA: University of Pittsburgh Press. This is a useful reader which includes some important readings on LGBT issues in Latin America.

de Santana Pinho, P. (2010) *Mama Africa: Reinventing Blackness in Bahia*. Durham, NC: Duke University Press. An analysis of the contradictory ways in which blackness is mobilized in Brazil.

Gutman, M. (ed.) (2003) *Changing Men and Masculinities in Latin America.* Durham, NC: Duke University Press. An edited collection that explores what it means to be a man in Latin America today.

Guzman Bouvard, M. (1994) *Revolutionizing Motherhood: The Mothers of the Plaza de Mayo.* Wilmington, DE: Scholarly Resources. An introduction to politicized motherhood through the story of the Madres of the Plaza de Mayo in Argentina's Dirty War.

Kampwirth, K. (2004) *Feminism and the Legacy of Revolution: Nicaragua, El Salvador, Chiapas.* Athens, OH: Ohio University Press. A book that explores the relationship between feminism and guerrilla warfare in three Latin American countries.

Lancaster, R. N. (1992) *Life is Hard: Machismo, Danger and the Intimacy of Power in Nicaragua.* Berkeley, CA: University of California Press. An innovative ethnography of the intersection between war, revolution and sexual identity in Nicaragua.

Puig, M. (1976) *El beso de la mujer araña/Kiss of the Spider Woman.* An Argentinian novel about the relationship in prison of a gay man and a political prisoner.

Wade, P. (1997) *Race and Ethnicity in Latin America.* London: Pluto Press. A very accessible theoretical overview of race and ethnicity in the continent.

Films

Conducta Impropia (1983), directed by Néstor Almendros and Orlando Jiménez Leal (Cuba). A documentary that deals with the persecution of homosexuals in Castro's Cuba from the start of the Cuban Revolution until the 1980s, including the labour camps to which gays were sent.

Doña Herlinda y Su Hijo (1985), directed by Jaime Humberto Hermosillo (Mexico). A Mexican movie that deals with the relationship between a mother and her gay son and his lover.

Las Madres de la Plaza de Mayo/The Mothers of Plaza de Mayo (1985), directed by Susana Blaustein Muñoz and Lourdes Portillo (Argentina). An Argentine documentary that deals with the struggles for justice waged by the Mothers of Plaza de Mayo during Argentina's Dirty War.

Strawberry and Chocolate (*Fresa y Chocolate*) (1994), directed by Tomás Gutiérrez Alea (Cuba). Set in 1970s Cuba, this film deals with the homophobia of that historical period.

No se lo digas a nadie (1998), directed by Francisco J. Lombardi (Peru). Questions of family relations, class, drugs and sexual identity are treated in this autobiographical account of gay Peruvian talk show host Jaime Bailey.

Poto Mitan: Women Pillars of the Global Economy (2009), directed by Renée Bergan and Mark Schuller (USA). A film that captures the suffering and the agency of a group of Haitian women workers.

Black in Latin America (2011), directed by Henry Louis Gates Jr (USA). A four-part documentary that deals with the African influence on Latin America and how legacies of colonialism and slavery have shaped the cultures of Haiti, Dominican Republic, Cuba, Brazil, Mexico and Peru.

Latin America's first gay wedding can be viewed at: www.youtube.com/watch?v= QcT9xWnE9eY

7 The politics of indigeneity

Learning outcomes

By the end of this chapter, the reader should:

- Be able to describe recent political mobilizations by indigenous groups and assess their significance
- Be able to evaluate how indigenous groups have adopted, adapted and resisted state and institutional spaces
- Have gained an insight into the specificities of indigenous struggles in a number of Latin America countries

Introduction

In Bolivia in 1990, 700 indigenous men and women walked 400 miles from Trinidad to La Paz in a "march for territory and dignity", a 35 day procession that "managed to shake up public opinion" and forced the Bolivian government into negotiations (Albó 1996). In the same year in Ecuador, a week-long national uprising organized by indigenous peoples paralysed the country's traffic and commerce. In 1992, Rigoberta Menchú, a 33-year-old K'iche Indian woman from Guatemala, won the Nobel Peace Prize, an event that just a decade earlier would have seemed unthinkable – in the early 1980s, as part of a brutal counter-insurgency strategy to eliminate guerrilla forces, Guatemala's indigenous peoples were being massacred by their own government with US military aid and the international mainstream media were all but ignoring their plight (see Box 7.1).

1992 was also the year in which indigenous movements across the Americas mobilized to celebrate 500 Years of Resistance, a counter campaign to the Spanish government's official celebration of the 500th anniversary of the "discovery" of America by Christopher Columbus, asserting that conquest, genocide and exploitation were not to be celebrated and producing in the process new levels of political connectedness and collaboration.

Less than two years later, in 1994, another unthinkable event happened, in front of the global media. The world was stunned on 1 January 1994, the date that the North American Free Trade Agreement (NAFTA) between Mexico, Canada and the United States came into force, when a group of masked indigenous guerrillas calling themselves the Zapatistas left their home in the Lacandón jungle to seize a number of cities and towns in Chiapas, Mexico, a struggle that gave way to a global network of solidarity with thousands of websites dedicated to the Zapatista cause.

In 2000, indigenous worldviews and protests were decisive in halting the privatization of water in Cochabamba, Bolivia (Laurie *et al.* 2002; see also Box 5.2). A few years later in 2003, a six-week uprising led largely by indigenous Bolivians would go on to bring down the president and in 2005, Bolivia elected an indigenous president explicitly committed to promoting indigenous rights through the nationalization of natural resources and land reform. While the Zapatista uprising was probably the most internationally visible indigenous mobilization in Latin America in the 1990s, the past two decades were witness to intense and inspiring modes of indigenous mobilization all over the continent. There is no doubt that the "return of the Indian" (Albó 1991; Wearne 1996) is posing one of the most substantive challenges to the development project in Latin America.

The political mobilization of indigenous peoples is not a new phenomenon. Indigenous peoples have resisted colonialism and the development project in both overt and covert ways for the past 500 years. But in more recent years indigenous mobilizations have become more politically visible and probably more politically effective. Indigenous peoples have protested with their bodies, by walking, occupying and blocking. They have got involved in truth commissions and national elections. Today, indigenous movements are increasingly transnational in scope, engaging with a range of national and international NGOs, government agencies and multilateral organizations such as the World Bank. Such transnational linkages, according to Andolina, Laurie and Radcliffe (2009: 1–2) have enabled them to move "from political obscurity to political centrality" (see also Brysk 2000). They have re-articulated dominant discourses of indigeneity to gain new modes of cultural and legal recognition and to create institutions through which indigenous identities and development concerns can be more effectively managed. Such policies have major implications for development thought, practice and policy. In other words, development is clearly being indigenized (Ramos, Guerreiro Osório and Pimenta 2009).

As we saw in Chapter 2, colonialism was devastating for indigenous peoples. Millions of Latin America's original inhabitants were killed or succumbed to abuse, exploitation and disease, and many others lost their land as a result of colonialism's desire to extract as much wealth as possible. The genocide and marginalization of indigenous peoples did not end after independence, with some national governments even believing that indigenous peoples should be eliminated. Others implemented policies such as *indigenismo*, which aimed to assimilate Indians into mainstream mestizo society through mechanisms such as Catholicism and education in the Spanish language. Mestizaje became a state ideology, in the sense that it was only by "conforming to a homogeneous mestizo cultural ideal" that one could enjoy the fruits of citizenship (Hale 2004: 16). As Hale (2004) writes, this was a unitary and individual model of citizenship that could not admit the culturally specific collective rights of indigenous groups. Even Nicaraguan revolutionary Augusto César Sandino hoped for the full integration of the Indians in a mestizo Nicaragua (de la Cadena and Starn 2007).

In the twentieth century, some Latin American governments adopted a corporate model in which indigenous peoples were encouraged to abandon indigenous identities in favour of class-based ones. Assimilationist ideas circulated alongside more contradictory discourses which saw indigenous peoples as a pure racial form in need of protection. Policies such as *indigenismo* met with varying degrees of resistance, including movements such as *indianismo* organized to resist the imposition of integration. As Quijano (2005: 58) writes, indigeneity can only be understood in relation to what he calls the "coloniality of the system of power". If power was not concentrated largely in the hands of European descended peoples, the term "indigenous" would have a very different set of meanings.

Development *qua* modernization and neoliberalism has also had terrible consequences for the cultural and territorial integrity of indigenous peoples. Land is central to indigenous cultures in the Americas, and embodies spiritual and cosmological dimensions as well as productive ones. The displacement of indigenous peoples from their ancestral lands through colonization programmes, the growth of agribusiness and plantation agriculture and economic activities such as oil extraction, logging and mining has been a common occurrence in many parts of the continent. In addition, indigenous peoples have had to confront the racist notion that their cultures are uncivilized, primitive and backward and doomed to disappear with modernization and development. In some countries, including Guatemala and Chile, indigenous populations have been seen as subversive and became the victims of campaigns of state-sponsored terror (see Box 7.1).

Box 7.1

Counter-insurgency and genocide in Guatemala

Geographical remoteness, political resistance and cultural resilience enabled the Maya people of Guatemala to survive and adapt to colonization. Although indigenous peoples make up the majority of the Guatemalan population and Guatemala did not undergo the kind of whitening policies or exaltation of mestizaje that took place elsewhere, ladino (mestizo) culture is dominant in Guatemala and the indigenous peoples are confronted with endemic racism and marginalization. Unequal land tenure means that landlessness and hunger have been persistent problems since the colonial period and have been the basis of Mayan resistance.

The lack of democracy (abruptly brought to an end by the CIA in 1954) and the appalling living conditions of indigenous peoples and low-income *ladinos* (*mestizos*) had given rise to diverse forms of political struggle. While some Guatemalans had formed a guerrilla army, the URNG (Guatemalan National Revolutionary Unity), and were mobilizing in the mountains, many indigenous plantation labourers had created the Committee for Peasant Unity (CUC) to defend their interests. It was not long after the creation of CUC that the indiscriminate killing of poor indigenous Guatemalans began, supported by large landowners, government-backed death squads and US military aid. The Guatemalan Civil War (1960–1996) lasted for 36 years and claimed thousands of lives, most of them Mayan indigenous people. The army killings were brutal and indiscriminate; those suspected of being enemies of the state were frequently tortured and mutilated, sometimes in horrific army-led massacres.

In 1980, a group of group of K'iche' and Ixil protestors who had occupied the Spanish embassy were burned to death in a police raid. Those killed included the father of Nobel Peace Prize winner Rigoberta Menchú (see Menchú and Burgos-Debray 1984). Subsequently, under the government of General José Ríos Montt and his scorched earth policy, the Guatemalan army massacred thousands of indigenous inhabitants and wiped more than 400 villages completely off the map. The government also forced young Mayan men to join the civil patrols and engage in anti-communist vigilance, pitting Mayans against Mayans. Refusal to participate could result in torture and even death.

A number of courageous human rights groups were formed by relatives of the victims to try to end the repression and bring the perpetrators to justice. The most important of these include the Mutual Support Group (GAM) (see Figure 7.1), the National Widows' Coordinating Committee (CONAVIGUA), the Council of Ethnic Communities "Runujel Junám" (CERJ) and the Communities of Population in Resistance (CPRs). The war continued until 1996, when peace accords were signed. A UN-sponsored truth commission estimated that 200,000 Guatemalans were killed in the conflict, 40,000 were disappeared and between 800,000 and 1 million Indians had been forcibly recruited to civil patrols (Drouin 2010), 83 per cent of victims identified were Maya and that the Guatemalan army committed 93 per cent of all atrocities (CEH 1999). More than a million Mayas fled the country as refugees or became internally displaced.

Although Guatemala's civil war was very much part of the Cold War counter-insurgency doctrine, most mainstream international media paid insufficient attention to the massacres there and concentrated largely on events in Nicaragua and El Salvador. While the truth commissions in Guatemala have established unequivocally that the state committed acts of genocide against its own people, bringing the perpetrators to justice has been extremely difficult. In 2012, however, former president General Ríos Montt went on trial for genocide and crimes against humanity.

Maya leaders were active in the 1996 peace accords, but lacked the level of organization necessary to articulate indigenous demands clearly at this time (Fischer 2004). In more recent years, as Postero and Zamosc (2004: 9) write, the pan-Maya movement has "grown in stature". In contrast to the situation in Bolivia, Ecuador and Bolivia, this growth has been achieved "without developing an overtly confrontational profile".

As noted in Chapter 6, the establishment of racial hierarchies was central to nation building in Latin America. While racial mixing between indigenous, African- and European-descended peoples was widespread, resilient racial categories were established in which skin colour and phenotype were decisive. The visible valorization of whiteness in all kinds of public spaces has been to the detriment of indigenous peoples, while the myth of racial democracy (see Chapter 6) perpetuated in some countries has made it difficult to respond effectively to racism. Furthermore, indigenous peoples are often commodified and exoticized by national governments as a tourist attraction, appearing in national dress in glossy official publications while in practice their rights are trampled on. In the 1990s the Guaraní people of Paraguay were seen as lazy and backward by the Paraguayan government, yet government officials responsible for promoting tourism would encourage tourists to visit Paraguay's picturesque Indian villages (Shankland 1993; see also Ramos 2006).

Although millions of them succumbed to acts of genocide, disease, exploitation and state-sponsored terror, the indigenous peoples of the Americas and their cultures did survive. Today, it is estimated that the indigenous population of Latin America is around 40 million, or 10 per cent of the population. Five Latin American countries are home to 90 per cent of the continent's indigenous peoples: Bolivia, Guatemala, Peru, Ecuador and Mexico (Yashar 2005). In two Latin American countries, Guatemala and Bolivia, the indigenous populations are in the majority.

Figure 7.1 *A protest held by the Group of Mutual Support (GAM) to protest at those kidnapped, tortured and killed by the military government, Guatemala*

Source: Philip Vine.

Peru and Mexico have indigenous populations in excess of 10 million. There are also large indigenous populations in Colombia, Brazil, Chile, Panama and Nicaragua and several indigenous languages including Guaraní, Aymara and Nahuatl have more than a million speakers. Between 8 and 9 million people across four countries – Guatemala, Mexico, Honduras and Belize – speak a Mayan language. In total, more than a thousand indigenous languages are spoken across the continent. As Rowe and Shelling (1991: 49) write, the conquest was catastrophic for Latin America's indigenous civilizations, yet "neither the colonial nor the republican regime has been able to expunge the memory of an Andean, Aztec and Mayan civilization" and these cultures have persisted.

Both indigeneity and indigenism are concepts widely used in the literature on indigenous peoples in Latin America, but lack a standard definition. In some cases, definitions of indigeneity are broadened to take in groups not previously identified as indigenous (for discussion see Ramos 1998; Niezen 2003; French 2009). In recent years, it has become more difficult to define who is and who isn't indigenous. Language, community residence, dress, diet, farming practices, mode of land ownership and class status are all significant but indigeneity cannot be reduced to any of these elements, as many indigenous peoples live in cities, speak Spanish and do not wear traditional dress – often in order to avoid discrimination and gain a degree of social mobility in racist societies (see Figures 7.2, 7.3 and 7.4). During the twentieth century, many indigenous peoples came to privilege a class-based identity as *campesinos* over an ethnic identity.

While Wearne (1994) believes that the most important element is probably self-identification, the state-led encouragement to assimilate and the social mobility that results from assimilation mean that many indigenous-descended people do not identify as indigenous (Yashar 2005). As indigeneity gets redefined in more empowering ways, a process of indigenous re-identification or re-Indianization is under way in some parts of the continent, with groups of people who had disavowed their indigenous heritage now reclaiming it (Field 1999; Warren 2001; Quijano 2005). It is useful to theorize indigeneity as a relational concept, as it is defined largely by relations with non-indigenous others including state actors. As de la Cadena and Starn (2007: 3) write:

> Reckoning with indigeneity demands recognizing it as a relational field of governance, subjectivities and knowledge that involves us all, indigenous and non-indigenous – in the making and remaking of its structures of power and imagination.

Figure 7.2 *Quechua woman and child, Sacred Valley, Peru*

Source: Quinet, Wikimedia Commons (Creative Commons Attribution 2.0 Generic).

Figure 7.3 *Aymara ceremony in Copacabana, Bolivia*

Figure 7.4 *Guaraní family, Mato Grosso do Sul, Brazil*

There are a number of factors that scholars believe help us to understand the intensification of indigenous political organizing and visibility in recent decades. Yashar's (2005) work, in particular, has attempted to explain the strength of indigenous organizing in Bolivia and Ecuador and its comparative weakness in Peru. Indigenous peoples have had to respond to negative processes that have affected their lands, livelihoods and cultural integrity, including colonization schemes and oil exploration. In Ecuador, oil exploration, initiated by Texaco in the 1960s, has had tragic consequences for indigenous health and environmental integrity and the magnitude of the destruction has galvanized indigenous opposition (see Sawyer 2004 and Chapter 5). Human rights abuses and the dismantling of social safety nets during the "lost decade" have also fomented an oppositional indigenous politics. In addition to responding to challenges, indigenous communities have taken advantage of varied spaces of opportunity to organize and express themselves politically.

In the mid-twentieth century, some Latin American governments were responding to growing political demands by their indigenous populations through a corporatist model. Corporatism is a form of "controlled mobilization" led by the state which attempts to control populations in return for certain rewards (Kampwirth 2002). While such regimes provided indigenous peoples with access to certain benefits such as agricultural credits, subsidies and other social programmes, they also constrained indigenous autonomy in significant ways.

Some indigenous groups did however create autonomous spaces within corporatist governance structures. Yashar believes that the weakness of the state and its inability to penetrate parts of the Andes or Mesoamerica facilitated the continuation of indigenous forms of governance and the projection of alternative ways of knowing and being. Kampwirth (2002) believes that the corporatist models of organization created by the PRI in Chiapas in Mexico had the unintended consequence of laying some of the groundwork for the subsequent Zapatista rebellion. At times, indigenous peoples organized themselves because states were simply too weak to impose their interests on indigenous communities or because these communities "converted the tools of control into organizational resources used against the colonizers" (Yashar 2005: 118). Democratization also brought new organizational possibilities to these communities. The return to civilian rule made it possible to express political views more openly. The end of civil wars and the signing of peace accords in Guatemala, Colombia and Peru have provided spaces for indigenous political engagement at the national level and an opportunity to articulate injustices.

Although democratization brought some political benefits and organizing possibilities, it coincided largely with the neoliberal economic reforms of the 1980s which brought hardship to low-income populations throughout the Americas, indigenous peoples included. Yashar (2005) sees this period as effecting a shift from a corporatist citizenship regime to a neoliberal one, and it was a shift that enhanced political rights while curtailing social ones.

Neoliberalism had dramatic and mostly disempowering impacts on indigenous populations. The social programmes that the corporatist model had brought were dismantled and export-oriented growth led to the expansion of cattle-ranching, mining, logging and oil exploration in and around indigenous communities, displacing indigenous families or polluting their land. Neoliberalism, with its emphasis on individualism, is anathema to the collective social and cultural systems that underpin indigenous cultures and its impacts have had a politicizing effect on indigenous groups, encouraging them to mobilize to defend their lands and their livelihoods.

Despite the hardships generated by neoliberalism, it also provided new opportunities for indigenous action. Hale (2004: 17) argues that some aspects of neoliberal democratization are compatible with indigenous cultural rights and, in particular, neoliberalism undermines mestizo nationalism, including the "distinction between the forward-looking mestizo and the backward Indian". Agencies such as the World Bank have embraced the idea of an empowered indigenous civil society able to take responsibility for its economic and political future.

Another factor that Yashar (2005) and Andolina *et al.* (2009) have characterized as central to indigenous mobilization has been the extent of translocal and transnational networks. Indigenous communities have long been connected to other entities, including peasant unions and the Catholic Church. In certain times and places, the Catholic Church, which has often acted to the great detriment of indigenous communities, has also come to vigorously defend indigenous rights (Martí i Puig 2010). A similar role has been played by the Protestant Moravian Church and its defence of Miskito rights on Nicaragua's Atlantic Coast (Hawley 1997; Dennis 2004). Some progressive church sectors influenced by liberation theology organized Christian Base Communities (CEBs), which are also an important component in the development of indigenous political movements.

Up to the 1980s, indigenous mobilization tended to be organized at the local or subnational scale and in many cases indigenous identities were subordinated to class-based ones (NACLA 1996). Their networks have become denser and more extensive in recent years, in part as a result of the development of new media technologies such as the Internet (see Chapter 8) which have enabled indigenous communities to communicate and collaborate with other indigenous peoples and with national and international NGOs – a kind of "scaling up" (Yashar 2005) or "jumping scale" (Andolina *et al.* 2009).

Indigenous groups in Latin America have been able to take advantage of multilateral initiatives to advance indigenous rights. In the 1920s, indigenous peoples were denied access to the League of Nations but now indigenous peoples are highly active within the UN and other bodies. A number of important initiatives have been established by the UN and other multilateral bodies to promote and protect indigenous rights. These include the creation of the Working Group on Indigenous Populations (WGIP) in 1982, the passing of Convention 169 by the International Labour Organization in 1989, the World Bank's Operational Directive to promote indigenous development in 1991, the establishment of the United Nations Permanent Forum on Indigenous Issues in 2000 and the approval of the UN Declaration on the Rights of Indigenous Peoples in 2007. All of these initiatives, designed to end discrimination against indigenous peoples, have strengthened the negotiating power that these groups possess when dealing with their own national governments. By developing linkages and solidarities across borders with other indigenous groups, NGOs and civil society activists, they have been able to develop new linkages and therefore new forms of struggle.

International collaborations have been fundamental in legal claims to land titling and the mapping projects that underpin them. For example, in the 1990s the Mayangna people of Awas Tingni in Nicaragua, with support from the World Wildlife Federation, the University of Iowa and a number of geographers and anthropologists with expertise in indigenous rights and in GIS and countermapping, turned to the OAS Human Rights Commission. They did so as a result of repeated incursions by transnational logging companies into their ancestral lands and concessions made by the Nicaraguan government to Dominican and Korean-owned logging companies (Anaya and Grossman 2002; Wainwright and Bryan 2009). The court ruled in favour of the Mayangna people and confirmed that Awas Tingni had a communal property right to their land. Consequently, the court ordered the Nicaraguan government to title this land. The court decision set an important legal precedent in the Americas because it asserted the right to collective, rather than individual, property ownership.

International collaborations of this kind means that after centuries of misunderstanding, discrimination and neglect within mestizo societies, indigenous politics are now centre stage and indigenous communities are both contributing to and benefiting from the globalization of indigeneity. Transnationalism can be a double-edged sword for indigenous peoples, providing new room for manoeuvre and sources of financial and symbolic support, but also potentially reproducing essentialized understandings of indigenous peoples (see Box 7.2).

Indigenous organizations across the Americas have many concerns in common. In particular, they are fighting for land rights and territorial autonomy, for the right to cultural difference, including the practice of indigenous customs and bilingual education, and for political

---

Box 7.2

Transnationalism and environmentalism

While transnational collaboration is a means for indigenous peoples to gain access to a wider sphere of influence, it is not without its risks and downsides. Environmentalism has been an important form of transnational solidarity between indigenous peoples and first world activists. As we saw in Chapter 5, international interest in environmental issues such as the deforestation of the Amazon rainforest grew rapidly.

Within such debates, northern environmentalists began to see indigenous peoples as the embodiment of sustainability. Indigenous peoples are often seen as being outside of history and as living simply in harmony with nature. Although this image is an oversimplification of indigenous lives, it took hold. When indigenous groups began to mobilize in the 1980s and 1990s, it was not for a vision of environmentalism as understood in the first world. Indeed, their fight was primarily for land rights, autonomy and control of resources.

As Sawyer (1997) documents, lowland Indians who had formed the Organization of Indigenous Peoples of Pastaza (OPIP) in order to fight for legal land ownership in the face of state-endorsed oil extraction began to couch their demands strategically in the language of first world environmentalism, and by so doing gained greater leverage in their negotiations with the Ecuadorian government. In press releases and interviews they began to mimic western apocalyptic understandings about the destruction of the planet, describing themselves as defenders of "the last frontier of uncontaminated jungle remaining in Ecuador" and as managers of "the lungs of the world and the patrimony of all living species on the planet" (Sawyer 1997: 71). This strategy forced the government into granting land titles to the protestors. By so doing, however, indigenous peoples reproduce stereotypical and essentializing understandings of themselves as noble savages and their societies as unchanging and ahistorical. The reinforcement of such understandings can perpetuate prejudices against indigenous peoples and thus complicate the demands for full citizenship.

participation and representation. These are all issues that complicate existing understandings of citizenship and call for their redefinition.

These common concerns notwithstanding, indigenous organizing in Latin America is quite heterogeneous and indigenous communities often have different objectives and different political capacities (Postero and Zamosc 2004; Yashar 2005). Indeed, concepts such as autonomy and citizenship can be articulated in different ways by different groups of indigenous peoples. Looking at the differences between Bolivia, Ecuador and Peru as well as at subnational organizing, in particular the difference between the highlands and the Amazon, Yashar (2005) finds substantial variations in relationships with the state, in terms of the possibilities of association and of escaping or being subject to state repression. While Bolivia, Ecuador, Guatemala and Mexico have all experienced high-profile indigenous protests, Peru has failed to develop strong forms of indigenous organization (Yashar 2005; Postero and Zamosc 2004). One possible explanation for this is that Peru's indigenous peoples were more "successfully" assimilated and many identify primarily with a campesino identity.

A survey of all the indigenous movements in Latin America is beyond the scope of this chapter and the next sections will provide more in-depth information about mobilizations in Mexico, Ecuador and Bolivia. Readers are encouraged to read also about indigenous politics elsewhere in Latin America, including important developments in Brazil, Chile, Nicaragua, Peru, Panama, Venezuela and Guatemala.

Mexico: the Zapatista rebellion

The Zapatista rebellion, which first captured global attention in 1994, has been one of the most widely studied indigenous mobilizations of the twentieth century. The Zapatistas have been referred to as Cyber-Indians and there are millions of websites dedicated to their cause. Indeed, the Zapatistas are a required component of any development studies or Latin American studies course, not only because of the significance of their struggle for development processes in southern Mexico, but also because of the ways in which they have inspired a global solidarity movement.

The Zapatistas emerged from Chiapas, in the south of Mexico. It was a region with a large indigenous population and ongoing tensions surrounding land tenure, racism and rural poverty. The pressure for land reform led the Mexican government to allow landless campesinos to settle in the Lacandón jungle, where they engaged in cattle ranching, coffee production and subsistence agriculture. In the aftermath of the 1980s debt crisis, the prices of these commodities fell, and existing inequalities were exacerbated. Liberation theologist and local bishop of San Cristóbal de las Casas, Samuel Ruiz, became a passionate critic of the abuses suffered by the indigenous peoples of Chiapas.

In July 1993, the Mexican army discovered guerrilla army training camps in the jungle. A few arrests were made but with only a few months to go until the North American Free Trade Agreement (NAFTA) came into force, the significance of this finding was played down. The Mexican government did not want Mexico to appear politically unstable to potential foreign investors. But on the day that NAFTA came into force, hundreds of armed Zapatistas wearing ski masks seized seven towns in Chiapas. Some had weapons; others had sticks carved into the shape of guns. It was a timely appearance as the inauguration of NAFTA meant global media attention was already focused on Mexico and the rebellion was able to draw attention to NAFTA's dark side, in particular how trade liberalization and deregulations were already undermining rural and indigenous livelihoods and food securities and how NAFTA would consolidate such hardship.

Unlike other revolutionary movements in the Americas, the Zapatistas did not aim to overthrow the state, but instead demanded "the democratic revitalization of Mexican civil and political society and autonomy for and recognition of indigenous culture" (Routledge 1998: 240). The Zapatista leader, Subcomandante Marcos, began to issue inspiring and charming communiqués, full of engaging political analysis, humour and references to indigenous cultures but also refusing any fixed definitions of Zapatista politics. They were uploaded to the Internet and were instrumental in garnering international solidarity for the Zapatista cause. Much to the chagrin of the Mexican government and the state-controlled media, Subcomandante Marcos very quickly became an international celebrity and was interviewed in a range of international media. Flusty (2004: 176) describes Subcomandante Marcos in Gramscian terms as an "organicized intellectual".

As a consequence of the power of their discourse, the Zapatistas' struggle has been described as a "postmodern revolution" (Dominguez 1998). As a weakly resourced army, the Zapatistas could easily have been annihilated by the Mexican army, but their superior discursive power has turned this struggle into a war of words rather than a war of bullets and in this respect, the Zapatistas have retained hegemony. The creative imagination of the Zapatistas repeatedly confounds and outwits the repressive measures of the Mexican state, rendering them politically ineffectual. In 2000, in a symbolic and politically effective gesture, it was announced that the Zapatista Air Force had attacked the barracks of the Federal Army with hundreds of planes (Lane 2003). The planes were however made of paper and each was loaded with a message or "discursive missile" to protest at the ongoing military incursions into Zapatista lands.

The Mexican army continues to repress the Zapatista movement, but international solidarity means it has to remain restrained in its use of repressive force. In the meantime, the Zapatistas have managed to secure their own autonomous spaces within Chiapas (see Figure 7.5). There are now Zapatista councils, known as *caracoles* within rebel territory, where health, education, land rights and women's rights are all actively promoted. Indeed, the Zapatistas, while drawing on well-established revolutionary traditions in Latin America, have also put forward new modes of development that are shaping the texture of indigenous political intervention throughout the continent.

The Zapatista struggle has produced political benefits for indigenous peoples elsewhere in Mexico. The governor of Oaxaca, out of fear that the Zapatista rebellion would take hold there, extended shared government to the indigenous groups in that state and reformed the electoral law to include indigenous modes of participation (*usos y costumbres* as opposed to party politics) (Esteva 2010). A few years later in 1998, a new law granting autonomy to indigenous communities was

Figure 7.5 *Support for the Zapatistas: a roadside store in Chiapas is painted with Zapatista images*

Source: Ond_ej _vá_ek, Wikimedia Commons (Creative Commons Attribution-Share Alike 3.0 Unported).

passed. As Esteva (2010) recognizes, the rights granted have however been frequently violated.

Ecuador: the creation of CONAIE

Indigenous peoples in Ecuador have had to fight to protect their ancestral lands from mining, logging, cattle ranching, African oil palm cultivation and oil exploration. CONAIE, or the Confederation of Indigenous Nationalities of Ecuador (see Figure 7.6), was formed in 1986 to protest at the national and multinational incursions into their ancestral lands and to convert Ecuador into a plurinational participatory democracy.

CONAIE was the result of the merging of two regional indigenous organizations to present a more unified struggle. Since the early 1990s it has organized large high-profile uprisings and demonstrations, blocking roads and occupying government buildings, which have forced the government into negotiations. These negotiations have produced important gains for indigenous groups in Ecuador, helping them to gain

Figure 7.6 *Members of CONAIE march in 2002 in opposition to the Free Trade Area of the Americas, Quito, Ecuador*

Source: Donovan and Scott, Wikimedia Commons (Creative Commons Attribution ShareAlike).

land rights and prevent further oil exploration. Increasingly, to gain more leverage, CONAIE has linked its struggles with those of international environmentalism (see Box 7.2), and has gained the support of international NGOs such as the Rainforest Action Network. As Perreault (2001) writes of indigenous mobilizations in Ecuador, indigenous struggles form what he calls "an identity/territory nexus" by which he means that they have brought together the fight for resources such as land with that for cultural identity and the call for a plurinational state. Scholars of indigenous movements in Latin America have characterized CONAIE as the most prominent and successful indigenous organization in Latin America (Zamosc 2004; Yashar 2005).

In 1996, CONAIE became involved in electoral politics. It created a political party, Pachakutik, and gained seats in congress. While involvement in electoral politics is fraught with complexities and its success in this arena has been uneven (see van Cott 2008), CONAIE is firmly established as one of Ecuador's most important political actors. Its demands have been both ethnic- and class-based. For CONAIE, citizenship encompasses on the one hand the right to be indigenous and practise indigenous customs and on the other to have access to land, water, fair prices and credit (Yashar 2005).

High-profile indigenous mobilizations in Ecuador should not however lead one to speak of the indigenous movement as if it were a coherent and homogeneous entity, as there are many differences between indigenous organizations, particularly when regional and community organizations are also taken into account (Perreault 2001). The Ecuadorian political landscape has changed dramatically in the past few years and indigenous demands are resonating with non-indigenous populations. As van Cott (2010: 389) states, Pachakutik came to appeal to "unions, leftist intellectuals, and citizens dissatisfied with the corruption, polarization and paralysis of Ecuadorian politics". Indigenous peoples constitute only 15 per cent of the electorate, but it is no longer possible for aspiring politicians to ignore the indigenous vote (Postero and Zamosc 2004).

Bolivia: claiming natural resources, indigenizing the spaces of politics

Lowland indigenous peoples first began to mobilize in Bolivia in the 1980s, creating the Confederation of Indigenous Peoples of Eastern Bolivia (CIDOB). Important mobilizations during the 1990s led to some positive political developments for indigenous populations, even though neoliberalism was generating economic exclusions and hardships. The constitution was amended to declare Bolivia a multiethnic and pluricultural nation and indigenous groups gained access to formal forms of political representation, at municipal level and also at national level within the newly created Secretariat of Ethnic, Gender and Generational Affairs (Postero 2004). By 2000, however, indigenous protests increased dramatically, particularly around natural resources such as water, gas and coca.

In addition to mobilizing against the privatization of water, indigenous Bolivians returned to the streets in 2003 after President Sánchez de Lozada announced a plan to export Bolivian natural gas to the US via a pipeline in Chile. Opposition to the plan was so great that it forced the resignation of the president. Indigenous peoples also mobilized over the right to cultivate coca and against the damage inflicted on indigenous livelihoods by US-funded coca eradication schemes. While coca is the key active ingredient in the production of cocaine, in its unprocessed form it is a mild stimulant and has been used by indigenous groups in the Andes for cultural, religious and medicinal purposes for centuries. Increasingly, Bolivians were demanding a new constituent assembly to create a new Bolivia, based on fair trade, cooperation, the protection of environmental resources and the right to a dignified livelihood (Beck 2006).

All of these demands coalesced in 2006 in the election of Bolivia's first ever indigenous president, Evo Morales. Morales is a native Aymara and previously was a coca grower and union leader. He is the leader of the Movement for Socialism (MAS), a political party that is closely aligned with the country's social movements and recent indigenous mobilizations. He won elections with decisive popular majorities in both 2006 and 2009. He presents quite a different personal style from other heads of state in Latin America, often preferring to wear his indigenous dress, a brightly coloured *chompa* or jumper made of alpaca wool rather than the conventional dark suit and tie (see Figure 7.7). Since coming to power, Morales has implemented agrarian reform and nationalized oil and natural gas. A close ally of Venezuela's Hugo Chávez, Morales has been a strong advocate of indigenous rights and has been vocally critical of neoliberal economic policies, the role of the industrialized countries in accelerating dangerous climate change and the attempt by the US to eradicate coca crops as part of the "war on drugs".

By 2009, as Taylor (2009) notes, things were clearly improving for ordinary Bolivians. Poverty and infant mortality levels fell, illiteracy was eradicated, a state pension scheme was introduced, and grants and scholarships have been provided for parents and students. Funds from the

Figure 7.7 Bolivian president Evo Morales and Minister of Culture Elizabeth Salguero attend a festival in Tarabuco, Bolivia

Source: Carwil, Wikimedia Commons (Creative Commons Attribution-Share Alike 3.0 Unported).

newly nationalized natural resources, which have found their way into social development programmes, have provided a massive stimulus to the Bolivian economy, allowing for budget surpluses and high economic growth rates. One of Morales' key achievements has been the promulgation of a new constitution in 2009, which puts indigenous rights centre stage. It establishes the Bolivian state as both plurinational and secular, it makes special provision for the traditional cultivation of coca, it limits land ownership to 5000 hectares, it establishes Spanish and 36 indigenous languages as official languages and it prohibits both the establishment of foreign military bases on Bolivian soil and the privatization of oil and natural gas.

In 2011, to further protect Bolivia's mineral deposits, enshrine Andean cosmologies into law and take a stand against global climate change, the Morales-led government passed a law in which nature was granted equal rights to humans based on respect for Pachamama, or Mother Earth. As de la Cadena (2010) writes, Andean indigenous groups have injected earth-beings into institutional politics in a way that disrupts conventional western understandings of humans and non-humans as distinct and separate. It is important to note, however, that Evo Morales has faced substantial challenges during his time in office, from middle-class secessionist movements as well as from indigenous groups opposed to a government project to build a highway through the Isiboro-Sécure Indigenous Territory and National Park (TIPNIS) (Achtenberg 2012).

Achievements and limitations

While substantial challenges for indigenous peoples remain and they are far from achieving the levels of social, economic and cultural development to which they aspire, indigenous mobilizations in Latin America have led to a dramatic shift in the terrain on which development is conducted. Many taken for granted assumptions about what constitutes good development have been challenged, as have dominant conceptualizations of nationhood and modernity.

Latin American nation-states are being reformed into plurinational entities and plurinationality is increasingly finding its way into Latin American constitutions. Indigenous peoples are making decisive contributions to the democratization of Latin American societies and to the contestation of neoliberalism. They have interrogated and broadened conventional understandings of citizenship by drawing attention to the crisis of accountability of systems of government and to the racism and assimilationist ideologies that underpin these systems. They have posited collective rather than individual rights. They have asserted indigenous people's right to exist on their own terms and have opened spaces in which indigenous identities and cultural practices can be expressed and claimed. Being indigenous is increasingly something to be proud of and indigenous peoples have become savvy political actors who are now making a difference in the corridors of power at home and overseas.

There are however important limitations to the new models of governance that are emerging. The endorsement of indigenous cultural rights by national governments and entities such as the World Bank, while offering important political opportunities, is also producing limitations to indigenous empowerment. Hale (2004) believes that a discursive binary has been drawn between what he calls the "indio permitido" (authorized Indian) and the unruly, rebellious and dysfunctional Indian or "indio insurrecto". Borrowing the phrase "indio permitido" from Bolivian sociologist Silvia Rivera Cusicanqui, who believes Latin American governments are "using cultural rights to divide and domesticate indigenous movements" (Hale 2004: 17), Hale states that indigenous mobilization is tolerated as long as it fits within the confines of neoliberal multiculturalism and does not challenge basic state prerogatives through conflict.

The dichotomy is also giving rise to divisions within indigenous communities, between those compliant with the system and who benefit from it and those who continue to be excluded. In Ecuador, Perreault (2001) observes a developing distance between an elite indigenous leadership that is well connected and embedded in a network of transnational organizations and those whose sphere of influence does not extend beyond the local community. Similarly, in Bolivia prior to the election of Evo Morales, Postero (2004) sees important limitations to what she calls state-led multiculturalism. She acknowledges that indigenous peoples have made important gains through their involvement in municipal politics, but she notes that they are forced to work within "prescribed scripted processes", which limits the demands they can make or the questions they can raise (p. 204). In other words, indigenous peoples can gain access to spaces of power, as long as they behave themselves and do not take their political demands too far and the western liberal framework is left intact. It is a framework that sits uneasily with both ethnicity and autonomy as understood by indigenous peoples.

Castillo (2006: 43) develops Hale's argument, stating that the presence of the dichotomy means that the Indian cannot be an autonomous subject but instead becomes a "marker for a certain kind of distanced and exotic collective". In the post-9/11 world, indigenous peoples who do not fit the authorized model are likely to be criminalized and even labelled as terrorists. In Ecuador, many indigenous leaders, even those who supported Rafael Correa's bid for the presidency, have been imprisoned, accused of "terrorism and sabotage". Those imprisoned include the presidents of two indigenous organizations, Marlon Santi and Delfin Tenesaca (Zibechi 2011). Similarly, in Chile an anti-terrorism law created by the Pinochet dictatorship is being used by the current government to arrest and convict Mapuche activists who are attempting to reclaim ancestral lands and in the process to discursively construct Mapuche as violent (Wadi 2011).

One of the interesting developments in Latin America that might disrupt the good Indian/bad Indian dichotomy is the emergence of a tendency for some Latin Americans to identify simultaneously as black and indigenous. Dominant culture sees black and indigenous as mutually exclusive categories (Hale 2004; Wade 2006; Anderson 2009). In part, this distinction gets discursively reproduced because of the ways in which blackness is associated with diaspora and uprootedness and indigeneity is associated with deep cultural roots (Anderson 2009). Yet in reality, Latin America is also populated by an Afro-indigenous population. On the Atlantic Coast of Nicaragua there are both Miskitos and Creoles, with the former adopting an indigenous identity and the latter a black one. In practice, however, the two groups are thoroughly mixed and the boundaries between them are highly fluid. People are encouraged to identify as either indigenous or Creole, yet there are many people who straddle such divides. Whether one identifies as Miskito or Creole depends in part on the community in which one lives and the discursive positioning with which one more closely identifies (Dennis 2004). While indigenous peoples can strategically draw on discourses of territoriality and understandings of themselves as pre-modern in order to gain cultural rights, blacks are able to identify with potentially empowering and resistant modes of style and consumption. In Honduras, the Garifuna people are practising what Anderson (2009) calls a "black indigeneity" or an "indigenous blackness" in which the racial distinctions between black and indigenous are refused or at least creatively negotiated. These developments are important not just for those who identify as Afro-indigenous but also for all racialized Latin Americans as they potentially constitute a means to overcome the discursive and material divides separating racialized populations and with them the ability to work together to forge an anti-racist politics.

Anti-essentialism and hybridity

The emergence of Afro-indigenous populations undermines the essentialism generally associated with racial formations in Latin America. As noted above, strategic essentialism has reaped some political rewards for indigenous peoples, particularly in terms of valorizing something that was previously devalued, but is also limiting in terms of claiming autonomy and reconfiguring nation-states along pluricultural or non-national lines. But in the long term, decolonization and the end of racism will need to also tackle the question of essentialism.

A starting point is to recognize that indigenous identities are dynamic and hybrid and their cultures, rather than being timeless and static, are in a state of flux. It is important to recognize that most indigenous peoples in Latin America have endured substantial contact with European society and have adopted, adapted and resisted European customs and ideas. The conquest and subsequent colonization brought an exchange of cultures.

Despite the brutality of colonialism, indigenous cultures have survived and continue to be passed on to subsequent generations. As Hale (2006: 112–113) writes, with respect to the Awas Tingni court ruling discussed above, the use of techniques such as participatory mapping to assist indigenous peoples "brings the essentialist–constructivist conundrum to the fore" in its attempt to manage the tension between the use of media technologies to construct a map and the emotional rootedness in place articulated by indigenous peoples. For Hale, this tension has political and cultural implications for our understanding of the legal victory.

It is important to recognize that indigenous peoples live in postcolonial societies and the historical fact of the conquest and contact with Europeans and other foreigners cannot be denied. Most indigenous peoples have of course been exposed to non-indigenous cultures all their lives. The idea that western forms of modernity merely impact on vulnerable indigenous groups, destroying their culture, overlooks the creative ways in which many indigenous peoples have long histories of cultural borrowing, exchange and indigenization of foreign ideas, material goods and technologies.

In the seventeenth century, the British presence on the Atlantic Coast of Nicaragua was viewed by the Miskito people as one of mutual benefit and reciprocity and it was as a result of contact with Europeans that the Miskito came to constitute themselves as an ethnic group (Helms 1971). In other words, their sense of strong cultural identity was forged because of rather than in spite of their contact with outsiders. For the Miskito people, the alliance with the British provided them with consumer goods and also helped to keep their common enemy, the Spaniards, at bay. Indeed, the Miskito developed what Hale (1992) refers to as an Anglo affinity, taking aspects of British culture that they found appealing, including the monarchical model, top hats and gold-braided uniforms, and they traded local goods for titles such as Lord Rodney or Lord Nelson (Dennis and Olien 1984). The embrace and indigenization of foreign elements continued in the nineteenth and twentieth centuries with the arrival of US companies in the region and furthered the development of a cosmopolitan consumerism based on cash incomes and imported cultural products. Just as many westerners feel drawn to indigenous cultures and values, there are aspects of western cultures that indigenous peoples find appealing. Both non-indigenous and indigenous people engage in selective borrowing of local and foreign elements. Being indigenous does not mean necessarily not having a cell phone or an iPod, but it might mean having these things and defending the right to cultivate coca or farm land collectively and communally.

We should then think of indigenous cultures as hybrid cultures, in which the contact and coexistence with Europeans cannot be denied. As Rowe and Shelling (1991) write, colonial culture was controlled through interpretation. While indigenous peoples were not able to escape colonial cultural forms such as Catholicism, they could blend them with their own traditions in ways that made them more palatable (see also de Certeau 1984). Latin America is replete with examples of indigenized and Africanized Catholicism, including the Day of the Dead in Mexico, Sihkru Tara in Nicaragua, Corpus Christi in Bolivia, the Iglesia Andina in Peru, Candomblé in Brazil, Winti in Suriname and Santería in Cuba.

Cultural exchange is not a one-way process. European cultures have also been hybridized and continue to be through the kinds of indigenous mobilizations discussed in this chapter. Indigenous epistemologies and ontologies have appeal for non-indigenous populations, particularly because of the way they resonate with broader oppositions to neoliberal globalization. If development was ever a thoroughly western invention, it is now an indigenized western invention.

Conclusions

While indigenous struggles across the continent have their own geographically and culturally specific features, they have a number of issues in common. The legacies of colonialism and the nature of nation building after independence have had similar outcomes for indigenous peoples in the sense that they have been exposed to a dominant belief that they were doomed to disappear or that they were backward and primitive and incapable of civilization and progress. Consequently, across Latin America, indigenous peoples have had to struggle for their right to exist and in particular for their rights to land, autonomy, languages and cultures, and in the process have had to confront a range of difficulties from outright genocide to attempted assimiliation and cooptation; compromise of ancestral lands through colonization or other means; deforestation and contamination of natural resources central to indigenous livelihoods; racism and limited social opportunities as a result of such racism; and the inability to protect linguistic and cultural integrity. In many places, indigenous peoples have been ignored and neglected; in others they have been actively discriminated against. Land rights, autonomy and cultural survival have therefore been central to indigenous organizing but are often understood in quite different ways by indigenous communities.

Summary

- The past three decades have seen a dramatic growth in indigenous mobilization.

- Indigenous cultures have been subject to genocide and discrimination but have survived colonialism and development.

- In recent years, indigenous groups have become more transnational in their political action.

- The Zapatista rebellion inspired a global solidarity movement.

- The Confederation of Indigenous Nationalities of Ecuador (CONAIE) has been highly visible in its struggles against neoliberal modes of development.

- In Bolivia, indigenous mobilizations have enabled the renationalization of natural resources.

- Indigenous peoples are enacting citizenship and indigenizing development in many creative ways but still face important cultural and political barriers and limitations.

Discussion questions

1. Why do some describe the events of 1492 as a discovery and others as a conquest? Which term do you prefer, and why?

2. Why have indigenous struggles developed more momentum and visibility in Ecuador and Bolivia than in Peru?

3. How did Guatemala's indigenous peoples get entangled in the counterinsurgency policies of the Guatemalan government in the 1980s?

4. What is the relationship between neoliberalism and indigenous mobilization?

5. How are indigenous movements becoming transnationalized? What are the advantages and disadvantages of transnationalization for indigenous struggles?

6. Outline the main dimensions of the Zapatista rebellion. Why do you think it has attracted so much interest from scholars and activists?

7. Why should we pay close attention to the convergence of black and indigenous struggles in Latin America?

Further reading

Andolina, R., Laurie, N. and Radcliffe, S. (2009) *Indigenous Development in the Andes: Culture, Power, and Transnationalism.* Durham, NC: Duke University Press. Based on in-depth fieldwork in Bolivia and Ecuador and focusing on relations between indigenous groups and other local, national and transnational entities, this book explores how indigenous peoples are reconfiguring development practice and the dilemmas created by indigenous engagements with development.

Menchú, R. and Burgos-Debray, E. (1984) *I, Rigoberta Menchú: An Indian Woman in Guatemala.* London: Verso. A classic (and controversial) text that tells the story of Rigoberta Menchú and the suffering and repression of the Mayan people of Guatemala.

Postero, N. G. and Zamosc, L. (eds) (2004) *The Struggle for Indigenous Rights in Latin America.* Eastbourne, UK: Sussex Academic Press. An edited collection examining indigenous movements in eight countries in Latin America, with an emphasis on struggles for rights and democratization and negotiations of neoliberalism.

Yashar, D. J. (2005) *Contesting Citizenship in Latin America: The Rise of Indigenous Movements and the Postliberal Challenge.* Cambridge: Cambridge University Press. A book that takes a comparative approach to indigenous mobilizations and attempts to illuminate the apparent differences in political effectiveness.

Films

When the Mountains Tremble (1983), directed by Pamela Yates and Newton Thomas Sigel (USA). A documentary that exposes the state-led terror carried out against the Mayan people in Guatemala. The filmmakers accompanied both the army and the guerrillas during the early 1980s.

The Spirit of Kuna Yala (1991), directed by Susan Todd and Andrew Young (USA). A film about the Kuna of Panama and their struggle to protect their homeland.

Encuentro with the Indigenous Peoples of Colombia (1997), directed by Robin Lloyd (USA). A delegation from the Colombian Human Rights Network visits the Cauca region and indigenous activists talk about the effect of colonization on their cultures.

The Real Thing: Coca, Democracy and Rebellion in Bolivia (2004), directed by Jim Sanders (Canada). A film that explores the impact of US foreign policy, including the "war on drugs" and the "war on terror", on the indigenous peoples of Bolivia.

We are the Indians (2004), directed by Philip Cox and Valeria Mapelman (UK/Argentina). A film about Argentina's last surviving Guarani Indians.

Bolivia – Narco State (2007), directed by Phil Cox (UK). A film on the cultivation of coca and the struggle against cocaine production under the government of Evo Morales in Bolivia.

La Voz Mapuche (*The Mapuche Voice*) (2008), directed by Pablo Fernández and Andrea Henriquez (Chile). A documentary that explores the struggles of the Mapuche people in Chile and Argentina to protect their cultures and identity and their ancestral lands from multinational companies.

Nuestra Historia está en la Tierra (*Our History is in the Earth*) (2008), directed by Eliezer Castro (Venezuela). A documentary that records the struggle by indigenous people in Venezuela to reclaim ancestral lands.

Hermana Constitución (*Sister Constitution*) (2008), directed by Soledad Domínguez (Bolivia). A film that looks at the creation of a Constituent Assembly in Bolivia and the drafting of a new constitution.

Indigenous Identity and Democracy in Mexico (2011), directed by Gabylu Lara (Mexico). A documentary that charts the struggle for political participation and representation by indigenous peoples in Mexico.

Websites

http://indigenouspeoplesissues.com
Indigenous Peoples Issues and Resources – news, articles and videos about indigenous peoples all around the world.

http://lanic.utexas.edu/la/region/indigenous
Latin American Network Information Center (LANIC) – links to information on indigenous peoples.

https://webspace.utexas.edu/hcleaver/www/zaps.html
A website maintained by Harry Cleaver at the University of Texas – contains links to Zapatista resources available on the Internet.

Communicating Latin American development

Media and popular culture

Learning outcomes

By the end of this chapter, the reader should:

- Be able to evaluate the centrality of media and popular culture to the development process
- Understand the diverse ways in which Latin Americans are consuming and producing media and their outcomes for development

Introduction

In the shanty towns that border most Latin American cities, it is not uncommon to find a television set or even a satellite dish in a makeshift home that has a dirt floor and no sanitation. Development practitioners in Latin America are often critical of this state of affairs, suggesting that the poor are somehow irresponsible for prioritizing frivolous television viewing when their children lack food, clothing or adequate shelter.

García Canclini (2001) urges us to approach this question differently. He believes that the media consumption of the poor is in fact highly complex and cannot be reduced to their location in an asymmetrical set of power relations in which powerful media corporations bent on global domination manipulate their irrational viewers and distract them from more important development needs. Instead, he shows how activities such as television viewing provide a sense of belonging, and as conventional spaces for democratic participation closed under neoliberal structural adjustment, they "can be good for thinking and acting in a meaningful way that renews social life" (2001: 47). In other words, media consumers can also be thought of as citizens and media consumption as the enactment of citizenship and a form of participation in the process of development.

Popular culture has always been central to Latin American development as a means of cultural expression, identity construction and political struggle. As the late Argentine sociologist Eduardo Archetti remarked in an interview, the creation of Argentine national identity owes more to tango, polo and football than it does to the export of meat and grain (Libendinsky 1999; see also Archetti 1999). Media, communication and popular culture more broadly constitute important vehicles through which people make the landscapes in which they live, contest relations of power, make sense of injustice and imagine a better life for themselves and their communities. Furthermore, much Latin American popular culture, including telenovelas, salsa and football, has burst on to the global scene, sometimes encouraging an interrogation of stereotypes about Latin America and Latin Americans in a way that has the potential to undermine neocolonial relations of power.

Despite this centrality, until recently popular culture was paid little attention in development studies, a neglect that began to change as a result of both globalization and technological change and the emergence of information and communication for development (ICT4D) as a specific area of concern for development studies. The so-called digital divide – the fact that some people do not have access to media and communication technologies – is increasingly seen as an obstacle to development and something to be overcome. Gaining access to the means of information and communication can be just as important as having access to clean water, decent housing, agricultural inputs, health care, education and democratic franchise, and indeed having communication media at one's disposal can also facilitate gaining these other more conventional development needs.

In terms of access to media, it is important to include both personal and grassroots media (such as owning a cell phone or participating in a community radio station) and mass information and entertainment media (such as being able to watch national news, CNN, Hollywood movies and telenovelas on television). Radio, Internet, television, cell phones and geospatial media technologies such as geographic information systems (GIS) are being used by ordinary Latin Americans to promote, engage with and contest the development project. There is still an important digital divide in Latin America, in terms of ownership of and access to television, cell phones and Internet, but it is one that is rapidly being overcome. While only 36 per cent of Latin Americans currently have access to the Internet compared with 78 per cent of North Americans, the rate of growth in Latin America is much higher. Latin America experienced a growth rate of 1037% between 2000 and 2011 compared with a rate of 151 per cent for North America (internetworldstats.com). There are substantial differences between Latin American countries. In Argentina, 66 per cent of the population is connected to the Internet, while fewer than 15 per cent of Bolivians, Cubans, Hondurans and Nicaraguans have access (internetworldstats.com).

The number of Facebook users in the region is also growing exponentially. By 2012, Brazil had 35 million, the fourth largest number of Facebook users per country in the world, and saw an increase of 300 per cent in 2011, the fastest rate of growth in the world (Burcher 2012). Cell phone ownership has undergone what can only be described as a boom. Mobile penetration in Latin America, at 89 per cent, is as high as the United States and higher than both India and China (Favell 2010). For many Latin Americans unable to afford a fixed landline or who are tired of waiting to receive one, a cell phone has been their first access to telephony. While the numbers of Latin Americans that own a cell phone and have access to the Internet are rapidly growing, some Latin Americans are still waiting for electricity.

Popular cultures, hybrid cultures

In order to develop a more nuanced understanding of contemporary Latin American media culture and its significance for development, it is worth thinking about popular culture more broadly. Just as Latin America's political struggles and economic models are central to development processes, the same can be said of Latin America's rich and diverse popular cultures. For del Sarto (2004: 160) "in Latin America the cultural dimension has become the paramount field of struggle for the resolution of socio-political and ideological conflicts", hence its centrality for Latin American development. Indeed, development can start to look quite different when you take popular culture seriously (for an example, see Cupples 2009).

Latin America's popular and media cultures are so rich and diverse that a single chapter can barely do them justice, but we can make some generalizations. First, although the experience of colonialism, imperialism and development as modernization has frequently proved devastating for mestizo, indigenous and Afro-Latino people, their cultures have not been wiped out through such processes. Indeed, as we saw in Chapters 6 and 7, they have shown extraordinary resilience and capacity for survival and creative hybridization. In 1940, Cuban scholar Fernando Ortiz coined the term "transculturation" as a means to think about this process. Rather than "acculturation", which sees European colonization as a one-way process through which indigenous cultures are modernized or westernized, transculturation looks at how European, indigenous and African peoples all exchanged properties and were all transformed in the process (Ortiz 1995[1940]); Rowe and Shelling 1991).

Even Catholicism as a cultural form imposed by the Spaniards did not survive intact in its new home; instead the indigenous people blended Catholic and pagan rituals, often making it appear to the colonizers that they embraced the Catholic religion, while they continued to worship their own deities. For example, the Mayan people of Santiago Atitlán in Guatemala worship San Simón, who is a blend of a Catholic saint and a pre-Columbian Mayan deity. Similarly, the Mexican Day of the Dead is a blend of the Catholic All Saints' Day with Aztec and Zapotec rituals to honour the dead (see Figure 8.1). Indeed, as French theorist Michel de Certeau (1984: 17), who worked in Brazil, Argentina and Chile, writes, subordinated peoples develop a range of techniques with which they "subvert the fatality of the established order". Moreover, this subversion is achieved by making use of the systems imposed by the Spanish which transformed the ways in which they functioned.

African slaves and their descendants enacted similar operations, and managed to keep African spiritual, religious and supernatural cultures

Figure 8.1 *Candy figures known as* **alfeñiques** *are used to celebrate the Day of the Dead in Mexico*

alive. They live on today in popular religions such as Santería in Cuba and Candomblé in Brazil. Such processes of cultural hybridization and resistance have been ubiquitous throughout the continent, and an analysis of popular cultures shows that indigenous and African traditions continue to be influential and in many cases are experiencing dramatic resurgence and growth. For example, the worship of Santa Muerte, or Saint Death, in Mexico City has grown dramatically in the past two decades. Santa Muerte appears to constitute a reworking of Mictlantecuhtli, the Aztec god of death, and on the first day of each month, thousands of devotees visit her shrine in the barrio of Tepito to leave candles or bottles of tequila in her honour (see Figure 8.2).

Intense ongoing processes of hybridization can be observed in Latin America's diverse and heterogeneous musical genres: not only in salsa, merengue and samba, but also in Andean rock, reggaeton, tecnocumbia and mangue beat. Music and other forms of popular culture are frequently used a mode of protest or as a means to contest established relations of power. Chilean Victor Jara (1932–1973), Afro-Colombian Joe Arroyo (1955–2011) and Argentine Mercedes Sosa (1935–2009) are among Latin America's most important "protest" singers whose fame far exceeds their national borders.

Figure 8.2 *Mexican followers of Santa Muerte visit a public shrine in Tepito, Mexico City*

Source: Thelmadatter, Wikimedia Commons (Creative Commons Attribution-Share Alike 3.0 Unported).

Latin America also has a well-established tradition of popular or political theatre, which draws inspiration from earlier traditions such as the Güegüence play (see Chapter 2). During the dictatorships, many Latin Americans developed forms of political, popular or testimonial theatre aimed at transforming political realities. The most well known and influential of these is probably Brazilian Augusto Boal's (1931–2009) Theatre of the Oppressed, created in the 1970s while Boal, who had been imprisoned and tortured in Brazil, was in exile. During the Pinochet dictatorship, Chileans used theatre to express opposition to the regime, although not without risk. In 1974, the members of Grupo Aleph were imprisoned after a satirical performance about the coup (Richards 2005). Latin America also developed an important Tercer Cine (Third Cinema, a play on the notion of Third World) movement in the 1960s, made up of politically committed film directors who belonged to the Cine de la Base in Argentina or Cinema Novo in Brazil, as a means to represent poverty and workers' struggles and contest the military dictatorships.

The developmental power of popular culture is demonstrated at the present time in the work being done in the favelas of Rio de Janeiro by civil society movement AfroReggae, which is using reggae, hip hop and other cultural practices to confront social exclusion and drug-related violence and in the process is bringing about dramatic and empowering social transformation. As Neate and Platt (2010) write, AfroReggae is an unconventional NGO made up of former drug traffickers who wish to get out of the drug trade and its culture of violence. It was formed in 1993 after a police-led massacre in a favela called Vigário Geral which left 21 people dead. Its work includes a reggae band, theatre, percussion workshops, dance, circus and the provision of computing facilities and the strategy includes sometimes blitzing poor communities with "cultural invasions" of music and dance that promote social and cultural interaction. As former drug trafficker and AfroReggae leader, JB, told Neate and Platt (2010: 68): "Before the arrival of AfroReggae in Vigário, the only thing you'd see about the neighbourhood was police brutality, drug trafficking and negativity. But now the news you see is all related to cultural projects produced by AfroReggae."

Cultural imperialism

Despite well-documented histories of vast cultural creativity and an ability to hybridize and indigenize foreign cultural elements amid the violence of colonialism, modernization or dictatorship, when it comes to contemporary media cultures there is an ongoing tendency to see Latin American cultures as vulnerable to mass media in general and to US cultural products in particular. For much of the twentieth century, and particularly when dependency theory was in its ascendancy, media and popular culture were understood largely as instruments of economic domination and cultural imperialism.

In an influential text entitled *How to Read Donald Duck: Imperialist Ideology in the Disney Comic*, written in Chile in 1971 by Ariel Dorfman and Armand Mattelart, the authors assert that Disney texts work ideologically to naturalize US American capitalism and imperialism. They argue that Disney's façade of innocence hides its ideological message, which is one that endorses and promotes US capitalist and consumerist cultural values. Disney comics also mobilize stereotypical understandings of race and gender and construct the third world as a site to be conquered and colonized (Dorfman and Mattelart 1975; Tomlinson 1991).

These understandings, while flawed, have proved to be resilient and constitute the basis of the cultural imperialism thesis. This thesis sees US American culture as all-powerful and able to eclipse and destroy the cultures of the periphery. It has however been convincingly challenged by scholarship in cultural studies which suggests that it is highly problematic to assume that a cultural product has ideological effects simply because it circulates in a given site. Such an assumption allows no scope for the fact that audiences can be quite heterogeneous in the ways in which they read and interpret texts. Even if a preferred ideological reading, such as the idea that we should all aspire to US consumer capitalism, can be identified, there is no guarantee that the readers of the text will produce that reading (see Tomlinson 1991; Barker 1989). While Latin Americans do consume many US cultural products and US media corporations clearly profit from Latin American media markets, it is now clear that the cultural imperialism thesis is simplistic and does not capture the complexity of media flows in conditions of globalization.

Over the past two decades, important work in cultural studies has provided more nuanced and hopeful perspectives on global media cultures, indicating on one hand that even global products become indigenized when they travel (Appadurai 1996), because of the ways in which people make sense of texts in culturally specific ways. A Hollywood movie would therefore be "read" differently in the highlands of Guatemala than it would in parts of the United States or indeed in downtown Montevideo.

Rowe and Shelling (1991) have identified three flawed theoretical assumptions in much of the literature on popular culture and their work overall is a particularly valuable response to the cultural imperialism thesis (Tatum 1994). It is frequently assumed, first, that local and peasant cultures are eroded by the mass media, second that Latin American societies are on an inevitable teleological trajectory to western modern culture, and third that the popular culture of subordinated peoples constitutes a Utopian space that automatically possesses the resources for imagining a different future. The first fails to apprehend the dynamic nature of culture and the ways in which the traditional and the modern are entangled, the second denies Latin Americans the agency to craft alternative non-western modernities, and the third assumes that the analyst has the capacity to see the positive in any cultural form.

Rowe and Shelling's (1991) work, along with that of scholars such as Jean Franco, Néstor García Canclini and Jesús Martín-Barbero, has interrogated these assumptions, emphasizing the activity rather than the passivity of the media consumer and showing how consumers of popular culture, rather than being manipulated by capitalist forces, are able to construct oppositional meanings out of popular media texts. Martín-Barbero (2004a, 2004b) urges us to pay attention to the ways in which audiences use and make sense of media, and the forms of hybridization or mediations that ensue. By mediations, he is urging us to focus specifically on the conditions of reception, because popular cultural memories and the practices of everyday life influence how media texts are received. For this endeavour, we need what he calls a nocturnal map so that we can "study domination, production and labour from the other side of the picture, the side of the cracks in domination, the consumption dimension of economy, and the pleasures of life" (Martín-Barbero 2004b: 311).

This approach reveals how media consumption is one of the ways in which people seek and make development. As Martín-Barbero suggests, such consumption "expresses just aspirations to a more human and respectful life" and should therefore be understood not as a frivolous waste of time or as evidence of capitalist manipulation but as "a form of protest and an expression of elemental rights" (Martín-Barbero 2004b: 312). His work has been central to the necessary dismantling of binary distinctions between the traditional and the modern, between popular culture and mass culture, and between the national and the transnational (del Sarto 2004), which inhibit the ways in which we theorize such cultural activity.

All forms of popular culture, whether produced by indigenous groups in rural communities or by multinational media corporations, are used by people in ways *that make sense to them* (see also Rowe and Shelling 1991). Media consumption should therefore more appropriately be understood as a site of struggle in which meanings are made rather than a site in which capitalist forces of domination are reproduced (Martín-Barbero 2004b). Even US cultural imports in Latin American fit into this category, as such "alien" products can, as Franco (2004: 184) asserts, "be used to resist or subvert authority".

Throughout Latin America, we find the ubiquitous juxtaposition of tradition and modernity (Rowe and Shelling 1991). Taking popular culture and its relationship with development seriously also enables us to see that Latin American development is not about modernity replacing or eclipsing tradition but about citizens and consumers being selective about which traditional elements they wish to retain and which modern elements they wish to adapt. Cultural consumption can be progressive, reactionary or contradictory. It is important therefore to avoid both technological determinism (the idea that technology has the same predictable impacts, positive or negative, everywhere) and social determinism (the idea that people are able to mobilize technology however they wish).

An alternative approach, which might be described as a relational approach or one informed by actor–network theory, sees humans and technologies interacting in networks that are composed of social and technological elements and mutually exchanging their properties. Agency then is a relational outcome of such interactions. We can approach development therefore as a process of cultural hybridization – ordinary people wish to protect and revitalize the cultural traditions they deem valuable, but some "modern" technologies might facilitate this.

It is important to recognize that popular cultural forms or media technologies do not have fixed meanings. They are not inherently good or bad, and the relational outcomes that they contribute to depend on the conjunctural and contextual factors that surround their use. Similarly, globalization has varied effects on popular culture: popular cultural forms can be weakened or strengthened through their insertion in processes of globalization. Consequently, it is valuable to understand the difference between what might be described as a political economy approach to the media that focuses on media ownership and a cultural studies approach that focuses on what it is that users and consumers do with media.

The cultural imperialism is also called into question through the phenomenon of reverse flows (Barker 1997). It is clear that Latin America is not merely on the receiving end of globalization, but is also a key contributor to processes of globalization. Although much US American music, film, television and Internet content finds its way into

Latin American homes, cinemas, bars, media devices and personal computers, it is important to recognize that Latin America exports culture along with coffee, gold and bananas. Cultural products now account for 6 per cent of Mexican GDP and in Colombia they account for 4 per cent, higher than coffee (García Canclini 2002). Salsa dancing is a popular pastime in London and Auckland, Bosnians and Russians are big fans of Latin American telenovelas, Gabriel García Márquez and Mario Vargas Llosa have been translated into dozens of languages, and for decades Latin America has produced some of the world's most famous footballers. The US mediascape is also increasingly dominated by latinos/latinas. Through their work, actors such as Salma Hayek, América Ferrera and Sofía Vergara are calling stereotypical representations of Latin Americans into question, sometimes through parody or comedy. Cultural globalization, as García Canclini (2002: 63, my translation) tells us, "is not a branch of genetic engineering, the aim of which is to reproduce in all countries clones of the American way of life".

Media in Latin America

The struggle for a free and independent media in Latin America has been a difficult one, and Latin America has often been a challenging place in which to be a journalist or independent broadcaster. Freedom of the press was particularly restricted during the military dictatorships in South America, in Cuba under Fidel Castro, in Mexico under the PRI and during the civil war in Colombia. Many journalists have been killed or disappeared for attempting to reveal forms of state repression or writing or broadcasting in defence of social or political justice. Throughout the continent, rulers "used an arsenal of means to guarantee a docile media, from direct censorship to arbitrary allocation of state advertising, from verbal intimidation to persecution of dissident journalists" (Waisbord 2002). The PRI, for example, controlled the media to such an extent that journalists were considered by some as little more than transcribers of official information (see Hughes and Gil 2004: 27). Low pay probably encouraged journalists to take bribes from politicians and businesspeople (Rockwell 2002).

Colombia's civil war turned the country into one of the most dangerous places in the world to be a journalist. As Lauría (2004: 36) writes, all of the warring factions in Colombia, including guerrillas, paramilitaries, the army and organized crime, have targeted journalists to prevent them from reporting on the war. At least 100 journalists have been killed as a direct result of their work in Colombia since 1980 (Greenslade 2006) and many more have engaged in practices of self-censorship. The number of journalists killed has thankfully fallen in Colombia in recent years but has risen elsewhere, especially in Mexico and post-coup Honduras. In 2012, Mexico tied with Pakistan to be nominated by the International Press Institute as the most dangerous place in the world be a journalist (Slovo 2012).

Like land and other economic sectors, we can observe substantial concentration of ownership in Latin America's media markets. Many of Latin America's newspapers, radio stations and television channels are owned by oligarchic, dynastic and patriarchal conglomerates that have frequently operated in support of state interests and authoritarian governments, complying with state censorship (Sinclair 2004). In some countries, the media are dominated by one company, one family or even one man. For example, Televisa in Mexico and the Globo group in Brazil are enormous conglomerates that have enjoyed substantial political support, while in countries such as Argentina, Colombia and Venezuela strong duopolies have developed (Waisbord 2002). At the same time, some competition to the likes of Globo and Televisa has emerged in the form of Grupo Abril and TV Azteca in Brazil and Mexico respectively, so these media markets too are becoming more duopolistic (Sinclair 2004). Similar concentrations of ownership can be observed in Central America. Rockwell and Janus (2002) describe how a Mexican media magnate, Angel González González, enjoys a virtual monopoly over commercial television in Guatemala and also owns three of Nicaragua's nine television channels.

Many Latin American conglomerates work in partnership with foreign companies and are expanding into new markets including the United States. There is substantial Spanish and US investment in Latin American media markets and Venezuelan channel Venevisión has its headquarters in Miami, the so-called "Hollywood of Latin America" (Sinclair 2004). Latin American media are not therefore only made in Latin America. Both Venevisión and Televisa are collaborating with Univisión, a US-based Spanish language television network. Univisión has turned into a media giant, and owns 62 television stations, 65 radio stations and three record labels, which all deliver media content in Spanish to the millions of Latinos who reside in the United States (Ballvé 2004).

The media has clearly become freer since the transition to democracy and the demise of the PRI. Waisbord (2000) has observed a substantial growth in critical and investigative journalism in the region in recent decades, which is having a disruptive effect on the traditional relationships between states, markets and media outlets. He asserts that watchdog journalism is on the rise and increasing numbers of journalists and newspapers are committed to exposing corruption, wrongdoing and other political scandals in a manner not seen in the past, and as a result are making important contributions to democratization. For example, the Colombian press revealed how Ernesto Samper used drug money to fund his electoral campaign – allegations that had important implications for his ensuing presidency. Rockwell (2002) reports how in the mid-1990s Panamanian journalists began to speak out against the bribery of reporters by politicians.

The press has become freer but media ownership has further consolidated in the wake of neoliberal policies that have encouraged privatization and liberalization (Fox and Waisbord 2002a). Telecommunications were central to the neoliberal attempt to bring in foreign direct investment. These policies have enabled already powerful media companies to gain even larger market shares, particularly in other media such as Internet service provision, and have created new possibilities for foreign ownership. At the same time, the globalization of media generally is creating new markets in Latin America for cable television channels such as CNN, which produces its own Spanish-language news channel, CNN en español, watched throughout Latin America.

A distinctively Latin American challenge to the likes of CNN has however been mounted by TeleSUR, a pan-Latin American television initiative led by Venezuela in partnership with Argentina, Bolivia, Colombia, Cuba, Ecuador, Nicaragua, Paraguay, Peru and Uruguay and broadcast throughout most of the Americas as well as in Europe and North Africa (mostly via satellite and via the Internet). While the concentration of media ownership is clearly of concern for both democratization and development more broadly (see the collection of essays in Fox and Waisbord 2002b), it is important not to paint an overly negative picture of its consequences. The concentration of media ownership is different from the concentration of land ownership, in the sense that unlike land you do not need to own the media in order to use it. As Tatum (1994: 206) states, drawing on Rowe and Shelling's (1991) work, "although the public cannot claim ownership of the media, it can exercise control over the meanings conveyed". Also, for-profit media corporations have to make programmes that are appealing to large audiences so they cannot just speak to elites.

Telenovelas

One of Latin America's most important cultural phenomena and a significant Latin American cultural export to the rest of the world is the telenovela. Latin American telenovelas are serialized forms of televised fiction that have some similarities to English-language soap operas but have their own continental origins in the Cuban radionovela and the Brazilian serialized newspaper literature known as *literatura de cordel*. They are often screened in prime time for six days a week, and most run for between three and six months and then come to an end. Unlike British or US soaps, there is definite narrative closure with "climactic, nation-paralyzing endings" (Soong 1999). They often command large budgets, advertising rates and audiences. Every dollar spent in production returns an average of $2.50 in advertising (Brown 2003). They are watched by men and women of all ages, they are frequently part of everyday discussions at home, at work and in the street, and the theme songs are often chart-topping hits. In rural communities, where many homes do not have television sets, it is not uncommon for large numbers of people to crowd in one house where there is a set at the same time every day to watch a popular telenovela. As Martín-Barbero (2004b: 321) remarks, the geography of television viewing matters (see Figure 8.3) as it "allows us to establish a symbolic topography of class usages" and he goes on to say that such neighbourhood relationships, which often appear anachronistic in late capitalism, are reproduced in Latin American communities around the telenovela and thus have an important social dimension.

Different countries produce different kinds of telenovelas. As López (1995: 261) writes:

> Mexican telenovelas are notorious for their weepiness, extraordinarily Manichean vision of the world, and lack of specific historical referents. At the opposite end of the spectrum, the Brazilian telenovelas are luxurious, exploit cinematic production values, and are considered more "realistic" for their depiction of ambiguous and divided characters in contemporary (or specific historical) Brazilian contexts. The Venezuelan and Colombian telenovelas lie between these two extremes, assuming certain characteristics and establishing their own differences. The Venezuelan telenovelas are like the Mexican in so far as they tend to privilege primal emotions over socio-historical context, but they substitute dialog and utterly spartan sets for the signifying baroqueness of the Mexicans' *mise-en-scène*. The Colombians, on the other hand, have followed the Brazilian model, making specific and pointed reference to the history and culture of the nation, although not by recourse to "realism," but through the use of an ironic/parodic mode that combines the melodramatic with comedy.

Figure 8.3 *A group of people catch up on their television viewing, San Cristóbal de las Casas, Mexico*

Source: Julie Cupples.

Telenovelas are often dismissed by elites as trash television that fuels mindless consumption and distracts people from more important political or economic issues. However, while many telenovelas rely on conventional melodramatic themes with plenty of interpersonal conflict, jealousy and betrayal, many others contain important social or political content. Furthermore, to return to Martín-Barbero (2004b: 324), it is important not to hastily dismiss the importance of melodrama for understanding the Latin American social condition as it constitutes "a fertile ground for studying the unmatched rhythms of historical development and *mestizajes* of which we are made . . . the melodrama taps into and stirs up a deep vein of collective cultural imagination".

The social impact of telenovelas became apparent at an early stage in their development. *Simplemente María*, a Peruvian telenovela that ran in the late 1960s, dealt with the theme of rural–urban migration. It told the story of María, a poor rural migrant to Lima who worked as a maid and a seamstress and attended literacy classes in the evening. It is a classic rags-to-riches story in which María eventually launches her own business. It was so popular that in a carnivalesque moment mixing real life and fiction (see Martín-Barbero 2004b: 327), thousands of Peruvians turned up to the filming of María's wedding at the Church of Santa Teresita de Niño Jesús in Lima, dressed in their finest clothes and carrying wedding gifts. As Singhal and Rogers (1999) write, María was a model of upward mobility for low-income working-class women and the show contributed to a substantial increase in both sales of Singer sewing machines and enrolment in adult literacy classes.

More recent telenovelas have addressed a wide range of development issues, including gender, racial politics and legacies of colonialism and slavery. For example, *Café con Aroma de Mujer* (Colombia) set in one of Colombia's coffee growing regions dealt with questions surrounding environmental politics, *Nada Personal* (Mexico) covered political assassinations and political corruption, *Los de Abajo* (Peru) was a telenovela about unemployment, *A Escrava Isaura* (Brazil) was a story about slavery and *Rei do Gado* (Brazil) engaged with the problematic of unequal land tenure. One Nicaraguan feminist NGO, Puntos de Encuentro, began to produce its own telenovela, *Sexto Sentido*, as a means to encourage young people to engage with questions around sexuality, unplanned pregnancy, family conflict, HIV/AIDS, drug and alcohol abuse and education, and to take control of these aspects of their lives (see Weinberg 2006; Howe 2008).

As López (1995) and others have noted, telenovelas have become important exports, especially for Brazil, Mexico, Colombia, Venezuela and Argentina. They have been hugely popular not only in other Latin American countries but also in Russia, Latvia, Poland, Bosnia, the Czech Republic, Angola, Indonesia and China. Brown (2003) states that most telenovelas pay for themselves in their home markets and then go to generate an average of $28,000 per programming hour when re-broadcast abroad. One Mexican telenovela, *Los Ricos También Lloran* (*The Rich Also Cry*), was so successful in post-communist Russia that the *Moscow Times* reported the streets emptying, workers downing tools and the temporary cessation of fighting on the Azerbaijani–Armenian border when the show aired (Martínez 2005). When Indonesians were gripped by Venezuelan telenovela *Kassandra*, the Indonesian government issued a warning that it was not acceptable to be absent from work in order to watch a telenovela (Soong 1999). The success and appeal of Latin American telenovelas have inspired many Eastern European and Asian countries to make their own.

Grassroots, community and indigenous media

The consolidation of media ownership is not the only important phenomenon shaping Latin American mediascapes. Just as corporate media entities are growing larger and more profitable, grassroots, alternative and community media initiatives are also flourishing. All over the continent, citizens are developing their own radio and television stations and using the Internet in innovative ways to gain access to information, communicate with others and make media. While many Latin Americans enjoy corporate media and occasionally use it in

Box 8.1

The Ugly Betty *phenomenon*

The most globally successful telenovela ever made is Colombia's *Yo Soy Betty la Fea* (*I am Betty, the Ugly One*), first broadcast in 1999 by RCN (Radio Cadena Nacional). It is the story of Betty, a well-educated, idealistic and "ugly" young woman who gains employment in a fashion company. The narrative deals with questions of female beauty and Betty's ultimately successful struggle to become "beautiful". It became a national obsession in Colombia, capturing half of the audience share (Akass and McCabe 2007).

It enjoyed similar success all over Latin America and among Latinos in the US, and audiences used Internet chat rooms to discuss the story and the characters (Rivero 2003). Rivero (2003) believes that Betty's success in Latin America can be attributed to the ways it created a space of contestation and debate around socially constructed notions of beauty, which in Latin America are related to discourses of race and whiteness. The fact that Betty was able to performatively transcend her status was possibly a source of empowerment for many Latin American women who have been on the receiving end of racial ideologies that valorize whiteness.

There have been international remakes of the show in Belgium, China, Croatia, Georgia, Germany, Greece, India, the Netherlands, the Philippines, Poland, Russia, Spain, Turkey and Vietnam. In the United States, ABC made a highly acclaimed comedy-drama version, *Ugly Betty*, which ran for four years from 2006 to 2010. In the US version the story shifts to New York, and Betty, the daughter of an illegal Mexican immigrant, works as a personal assistant at fashion magazine *Mode*. The narrative moves between the competitive world of *Mode* in downtown New York and Betty's family life in Queen's, living with her father, sister and nephew. While the US version dealt with themes of beauty and appearance, it also raised important issues for the many US-based Latinos, whose parents may have migrated illegally to the United States. *Ugly Betty* received praise from Californian congresswoman Hilda Solís for its positive portrayal of Latino life in the United States (Katzew 2011).

resistant and subversive ways, many others want to make their own media to report on local perspectives and events, promote development, resist oppression, forge a sense of belonging, construct cultural identities and fight for human rights and dignity.

Latin America has a long historical tradition of participatory community radio, dating back to the 1940s. Community radio is a grassroots radio, created by and for local communities, to provide information and entertainment, to raise awareness of local events and local concerns and to foment political transformation and revolution. Early examples include Radio Sutatenza and the ACPO (Cultural Popular Action) broadcasting school in Colombia, the participatory radio stations established and operated by the Bolivian tin miners to make their political demands heard, and the radio stations used by the revolutionary Cuban government to promote mobilizations around health and education (Beltrán S. 2004).

During the civil war in El Salvador in the 1980s, the FMLN guerrillas ran their own underground radio station, Radio Venceremos. It began at a time of extreme media censorship when you could be arrested for handing out a political leaflet. It broadcast clandestinely from Morazán for 12 years, constantly challenging the regime of terror being waged by the Salvadoran government with a mixture of commentary, revolutionary music and political information (López Vigil 1995). Colombian guerrilla army the FARC ran a radio station for 15 years, until it was shut down by the Colombian army in 2012.

The Atlantic Coast of Nicaragua has developed a strong tradition of participatory radio. Many local stations including Radio Caribe and Kabu Yula in Bilwi have open-mike policies, whereby ordinary people can go into the radio station and are given air time to express their views. People often use these opportunities to communicate with local politicians, particularly when they feel that other efforts to gain political redress have failed (Glynn and Cupples 2011). Community radio stations have also flourished in Venezuela since the election of Hugo Chávez. The private media are largely opposed to Chávez and it became apparent during the attempted coup in 2002 that they were presenting a very partisan and incomplete account of events, falsely reporting for example that Chávez had resigned and that Chávez supporters had killed protestors (Wilpert 2004; Fernandes 2010). Consequently, ordinary citizens had to find alternative ways to distribute information about what was happening. After the coup, the Chávez government recognized the importance of community media, passing legislation and providing state funding to facilitate its activity. According to Fernandes (2010), there were 13 licensed community radio and television stations in 2002. By 2007, this had increased to 193 licensed stations and around 300 unlicensed ones.

Scholars of community media have emphasized how they emerge in a context in which people perceive limitations or lack trust in corporate mass media. They constitute important ways in which ordinary people construct place-based citizenship and thus contribute to the building of a public sphere (Winacur 2002; Remedi 2004; Fernandes 2010). For Remedi (2004), community radio stations democratize access to the means of cultural production and enable people to address marginalization and make political claims.

Winacur's (2002) work on the reception of community radio in Mexican homes emphasizes how through community radio, the home becomes a site in which people reflect on the quality of public services and on the everyday lived experience of poverty and underconsumption in their own local neighbourhood. In Venezuela, community media provide an important and self-affirming revalorization of everyday life in the barrio. As Fernandes (2010) observes, whereas the commercial media tend to emphasize the negative aspects of barrio life, such as crime and violence, community media emphasize positive cultural activities and everyday life. The communication of popular localized knowledges that contest and rearticulate official narratives enables popular communicators to begin to transform "stigmas of place and culture into positive signifiers of collectivity and community" (Fernandes 2010: 175). Fernandes also shows how community radio is a site in which people are able to engage in debates about race, cultural difference and indigenous histories of resistance. Such discussions might help people to begin to feel differently about their lives and the places in which they live.

One politically innovative use of radio was that enacted by the APPO (The Popular Assembly of the Peoples of Oaxaca) in Oaxaca, Mexico, in the wake of a teachers' strike in 2006 that was repressed by state forces. The harnessing of media to contest injustice was successfully documented in the film *A Little Bit of So Much Truth*. The teachers used their radio station, Radio Plantón, to communicate the need for the strike, highlighting in particular the low salaries paid to teachers and the fact that too many school children were forced to attend school hungry and without books or shoes. Before long, the teachers' strike turned into a popular rebellion involving indigenous groups, women's movements, environmentalists, trade union members and human rights advocates in which massive discontent with the PRI governor of Oaxaca, Ulises Ruiz Ortiz, erupted. For many Oaxacans, he represented the electoral fraud, corruption and social neglect that had become endemic to PRI leadership throughout Mexico, and this popular struggle led to the creation of the APPO. The radio communication had a mobilizing effect on the population, who came out in support of the movement, demanding the resignation of the governor.

When Radio Plantón was destroyed in a police attack, the movement began broadcasting from Radio Universidad instead and responded with the mass occupation of government buildings and the erection of barricades throughout the city to prevent further repression. The government also destroyed the transmitter of Radio Universidad and so the movement responded by taking over 12 commercial radio stations as well as television station Canal 9. Events in Oaxaca demonstrate the power of community media to bring about political change. While the

APPO did not succeed in toppling the governor, the media mobilization has had a long-term transformative effect on the city and state of Oaxaca (Esteva 2010).

Trigona (2004) describes how the *piquetero* or unemployed workers movement launched its own television station in the wake of the collapse of the Argentine economy in 2001. The mainstream media had criminalized the movement so its members felt they needed their own media outlet through which they could explain their political action and the motivations that underpinned it in their own words. With the support of the Grupo Alavío video collective, they produced a series of powerful reports that were watched by local people who began to engage with the issues in new ways.

In 2006, Grupo Alavío launched Agora TV, a community television production project that aims to provide an alternative mediaspace for working people and grassroots activists. It broadcasts over the Internet and so anybody with access to the Internet can log on and watch videos about labour and other social movement struggles. Grupo Alavío is also providing film-making skills to workers and participants in social movements to enable them to make their own videos and tell their own stories (Trigona 2007).

The form of grassroots media that most calls conventional binaries of tradition and modernity into question is the use of media by indigenous groups. While western media has often been viewed with suspicion by indigenous peoples, as they have for such a long time been subject to the colonizing gaze of first world photographers and filmmakers who are there to document a culture assumed to be on the brink of extinction (Ginsburg 2002), in recent decades it has become clear that indigenous media, particularly videomaking, has become "a crucial technology for the reinvention of indigenous cultures, directed against the long-standing discrimination of indigenous people, languages, medicinal practices, as well as their social and economic relations" (Schiwy 2003: 116). As Salazar (2009: 511) remarks with respect to indigenous media in Bolivia, indigenous media makers "have stressed the importance of appropriating media for decolonising the intellect".

Indigenous media generally combine technological modernization with the defence of traditional cultures and practices. It is clear that western media technologies when controlled by indigenous media makers can be put to alternative uses, and they are proving to be a useful tool for telling stories using indigenous peoples' own words and images and for protecting and revitalizing indigenous languages and cultural traditions.

A number of indigenous groups throughout Latin America have become active producers of their own media. Anthropologist Terence Turner, who has worked with Kayapo media makers in the Amazon for many years, has shown how they have creatively and effectively appropriated western media technologies in pursuit of their own cultural and political purposes. These purposes include the preservation and revival of traditional tribal practices, opposition to government policies that harm their interests and connection with other indigenous groups in the Amazon and beyond (Turner 1991, 1992, 2002). The juxtaposition of tradition and modernity can be seen in the images Turner includes in his work of Kayapo in traditional tribal dress with video cameras slung over their shoulders (see Turner 1992).

The indigenous use of media demonstrates how culture is not automatically swamped or destroyed by dominant and western media technologies. If used appropriately, modern media technologies are powerful tools of cultural preservation and assertion and can be used to forge alternative indigenous modernities (see Salazar and Córdova 2008: 52).

Making indigenous television and video is not easy. Most indigenous groups struggle to access the funding they need to purchase equipment. Some have overcome such difficulties, producing film and television in native languages that provides an indigenous perspective on the world and deals with themes often sidelined in mainstream media, such as land rights, environmental pollution, indigenous cultural celebrations and festivals, traditional medicine and bilingual education (see Figure 8.4). Latin America indigenous media makers have found that collaboration with groups elsewhere in the world is one way to overcome the funding constraints and forms of social and economic marginalization that they experience within the nation states in which their territories are located. Such collaboration enables an exchange of strategies and an exchange of content, so that indigenous television channels can show films made by indigenous groups elsewhere in the world. For example, a collaboration known as ¡VIVA! brings together indigenous media makers in Nicaragua with community artists in Panama, Mexico, the United States and Canada to produce a dynamic and reciprocal reflection on decolonization and social justice in the Americas (Glynn and Cupples 2011; Barndt 2011). While video has been a powerful tool for many indigenous groups, others, such as the Zapatistas (see below), are also making good use of the Internet, to connect with other indigenous people and supporters around the globe, demonstrating that there is "no necessary contradiction between technological modernization and grassroots mobilization" (Yúdice 2003: 106; see also Halleck 1994).

Figure 8.4 *Canal 7, an indigenous community television channel broadcasting from Nicaragua's Atlantic Coast*

Source: Kevin Glynn.

In recent years, many local communities and indigenous groups have been using geospatial media technologies such as GIS and global positioning systems (GPS). Maps have of course been important instruments of colonial control (Barton 1997), but there is evidence that they can be put to more decolonizing ends. Community mapping projects, often also referred to as countermapping or participatory mapping, are being used to map territories, often with a view to securing legal title and preventing the loss of ancestral lands to colonization schemes or logging projects.

In the 1990s, the late geographer Bernard Nietschmann participated in a community mapping collaboration between geographers, marine biologists and Miskito environmentalists and lobster divers to map the Miskito Reefs and surrounding waters. These waters contain valuable resources and therefore are constantly illegally invaded by lobster pirates and international fishing fleets. The team used GIS, GPS and Hi-8 video to record the indigenous knowledges of the Miskito team members and produce an inventory of the area. These maps were then used to seek international conservation support and to monitor the marine environment (Nietschmann 1994). As Nietschmann (1994) argues, maps are powerful tools, and just as maps have been used to conquer and colonize territories, they can also be used to defend and reclaim territory.

Since then, community mapping projects have proliferated throughout Latin America. Shipibo communities living in Cantagallo in central Lima, Peru are developing their own community maps to assist their ongoing negotiations for legal title with the municipal authorities and reduce conflicts with other groups living in the same community (see grassrootsmapping.org). While countermapping began with indigenous peoples, it is now being used by different groups of people. For example, a grassroots community mapping project in collaboration with the Center for Civic Media at MIT known as Crónicas de Héroes (Chronicles of Heroes) developed in Ciudad Juárez as a means to map examples of positive civic action that happen daily in the city in order to challenge the overwhelmingly negative view of Ciudad Juárez as a place of drug-related violence. People go online and map acts of kindness or help they receive from strangers that often go unnoticed. The project has now spread to Monterrey and Tijuana (see http://cronicasdeheroes.mx).

It is important however not to over-celebrate the value of geospatial technologies. As Wainwright and Bryan (2009) convincingly argue, drawing on mapping projects in Belize and Nicaragua, the dynamics of map production are fraught with complexities and can have undesirable outcomes for indigenous groups. The act of mapping might for example reconfigure indigenous land as property and thus merely rework rather than undermine colonial or neoliberal relations of power. To be used for example in a courtroom, the map has to be legible to western experts and must therefore adhere to western cartographic norms (see also Bryan 2009).

The close relationship between participatory mapping and military goals poses ethical complexities for indigenous and other marginalized peoples. In 2009, the American Geographical Society and the US Army's Foreign Military Studies Office (FMSO) created a mapping project, México Indígena, which aimed to map indigenous land in Mexico in order to build a GIS database. When indigenous groups in Oaxaca learned of the military involvement in the project, an intense controversy erupted (see Bryan 2010; Cruz 2010).

The Internet

There is no doubt that the Internet and the Web 2.0 environment more broadly have had a transformative effect on everyday development geographies in Latin America, enabling people to "jump scale" (Smith 2001; Warf 2008) and transcend the confinements of their own locales. The Zapatista rebellion, in which partially armed and masked guerrillas took over a number of towns and cities in Chiapas on New Year's Day 1994, the day that NAFTA came into force, was the first and one of the most important political struggles in Latin America to be fought in cyberspace as well as in the towns and jungles of Chiapas. The rebellion was well timed to take advantage of the global media attention focused on Mexico because of NAFTA. An innovative use of Internet and other forms of global media created massive transnational solidarity with the Zapatistas.

In the 1990s hundreds of websites dedicated to the Zapatista cause appeared and spread an anti-NAFTA, anti-neoliberal and pro-indigenous message across the world. As Froehling (1997) notes, this creative use of the Internet is somewhat surprising as the vast majority of Zapatistas do not even have access to electricity, let alone an Internet connection. The information on the Internet was not however distributed by the Zapatistas themselves from a central command site but developed in and replicated on many different sites around the world.

While their communications are rooted in the local realities of Chiapas and the forms of poverty and exclusion that underpin the rebellion, they have resonated with global struggles throughout the world. As Froehling (1997: 298) writes, "The causes of an indigenous movement, women's rights, anti-NAFTA sentiment and rhetoric against neoliberalism . . . provided a set of powerful rallying points". The Zapatistas have inspired a number of global struggles around the world, including the opposition to the WTO in Seattle in 1999, Iran's Twitter Revolution and the Occupy Wall Street movement.

While the Zapatistas are probably the most prominent and well-researched example of the use of the Internet to imagine an alternative development trajectory, Latin America abounds with examples of creative Internet use that show how the terrain of political struggle is shifting through its mediation (see for example Box 8.2). Pachico (2009) believes that a "quiet online revolution" is taking place in rural and isolated communities in Latin America. She describes how violence perpetrated by the Colombian army against protestors was filmed with a cameraphone and then quickly uploaded to the Internet. Similarly, residents of Tacueyó, an indigenous reservation in Colombia, used the Internet to denounce an attack and takeover of their community by FARC guerrillas, enabling them to amass crucial international solidarity and support very quickly.

In 2007, a peaceful protest against water privatization in Suchitoto, El Salvador, was brutally repressed by state forces and 14 protestors were charged with acts of terrorism. While such political violence is not uncommon in El Salvador, on this occasion the violence was filmed on cell phones and uploaded to YouTube. The images were spread rapidly around the world, and shared among activists and solidarity committees. In 2008, the charges against the protestors were dropped and there is no doubt that global media visibility played an important role in this outcome.

It is clear that new media provide an alternative avenue through which protests can be registered. During the 2007 Costa Rican referendum on CAFTA, the Patriotic Movemement for No (those opposed to the free trade agreement) found itself at a distinct disadvantage *vis-à-vis* the Yes campaign, which had the financial backing of the Costa Rican and US governments along with the national and multinational business sector and the support of both state and private television channels and Costa Rica's main daily newspapers, *La Nación* and *El Día*, and could therefore mount a multimillion dollar advertising campaign in the mainstream media. A leaked government memo also revealed an intention to use the mainstream media to incite fear of job losses and political instability should CAFTA not be approved.

The No movement responded to the unequal struggle by transforming the urban landscape of San José and other cities and towns with No T-shirts, badges, leaflets, banners, posters, graffiti, bumper stickers, street theatre and marches and by occupying alternative media sites, especially radio and the Internet. The government and a number of businesses tried to silence the radio stations carrying the opinions and voices of No activists by removing their advertising. In spite of this, anti-CAFTA radio programmes such as *Mujeres del No* (*Women of the No*), *Sepamos Ser Libres* (*Let's Learn to Be Free*), *La Semilla* (*The Seed*) and *Del Pensamiento a la Acción* (*From Thought to Action*) survived and were hugely popular (Cupples and Larios 2010).

During a severe electricity crisis in 2006 and 2007, those opposed to the privatization of electricity in Nicaragua and the perceived incompetence of Spanish multinational and electricity distributor Unión Fenosa engaged in creative forms of mediated activism. For example, consumer advocacy groups have made use of media spaces, especially radio, to denounce Unión Fenosa and inform people of their rights as consumers and citizens. Critiques of Unión Fenosa proliferate across diverse media spaces. The Internet is full of information about its alleged abuse, including a web-based campaign to rename the company Unión Penosa (Shameful Union) and a website, killfenosa.org, that existed for a while and enabled consumers to post comments on their experiences dealing with the multinational. Consumers also posted large numbers of comments on online newspapers on articles on the electricity crisis, and a video appeared on YouTube in which Unión Fenosa appears in a Managua barrio as Darth Vader, who represents the dark side, and is defeated by a local man armed with a machete (Cupples 2011).

The new media ecology has also been invaluable for gay activists in Latin America. As Encarnación (2011: 110) writes:

> Gay activists have also been exploiting new media to create a gay cyberspace of almost boundless benefits. In countries such as Argentina and Colombia, where the Internet reaches almost half the population, being gay has become a less isolating experience. The comparatively low cost of email, blogs, websites, and social networks has allowed a multitude of gay organizations to develop an online presence that depends for its existence on relatively few material and human resources and that affords the gay community a wealth of services – from archives of gay history to AIDS counseling to chat rooms in which gays can interact with each other without fear of being ostracized or attacked. The Internet has also functioned as a postbureaucratic universe of interest representation that allows gay activists to get their message across when "old" media outlets such as television and newspapers prove reluctant to do so.

While many Latin Americans still do not have access to the Internet, it is dramatically changing the ways in which political struggles for development are waged. As Kowal (2002: 110) writes, its development potential "is magnified by the democratic nature of the technology with its ability to disseminate information to a global audience almost instantaneously to accommodate an unlimited number of servers and users".

Conclusions

No study of Latin American development can be complete without paying attention to the role played by media and communication technologies. Consuming and making media are part of the ways in

Box 8.2

Yoani Sánchez and the Cuban blogosphere

The new media environment provides opportunities for Latin Americans to overcome press censorship and restrictions on freedom of speech. Access to the Internet is limited in Cuba, but nonetheless a Cuban blogosphere or blogostroika (Venegas 2010) has emerged. Portals such as desdecuba.com are providing a site for alternative media expression and citizen journalism. One Cuban blogger, Yoani Sánchez, has been writing a regular blog, *Generation Y* (desdecuba.com/generaciony), about everyday life in Cuba since 2007. She also posts regularly to social networking sites such as Twitter and Facebook and her story has been published as a book (Sánchez 2011). Her posts reveal the struggles faced by ordinary Cubans, including those of political expression and restrictions on the ability to travel and access the Internet and other media technologies such as cell phones. Occasionally, her ability to blog has been restricted and she has had to send her blogs via email to friends in other parts of the world who then post them on the Internet and email comments to her. One 2009 post described a temporary detention and beating at the hands of Cuban state forces (Hamilton 2009).

As Venegas (2010: 176) writes, in a context of high political dissatisfaction, digital technology should not be seen as "a luxury but as a potential avenue for self-realization outside the parameters demarcated by state bureaucratic mechanisms". Sánchez's blog has had a huge global impact and receives about 1 million hits a month (Hamilton 2009). It is translated into 17 languages and she has won numerous international awards for her work. *Time* magazine included her in the list of the 100 most influential people in 2008. She also featured in *Foreign Policy* magazine's list of the top global thinkers of 2011.

Despite international acclaim, Sánchez's work does not go uncriticized. Scholar Salim Lamrani has for example questioned her apparent close links with western embassies, her unwillingness to acknowledge Cuba's considerable social achievements in areas such as health care and education, and the veracity of an alleged attack by state forces (see Lamrani 2009, 2010).

which people make and contest development. Many people struggle to gain access to media in the same way they struggle for the right to vote without intimidation or for clean water or electricity, and these struggles need to be taken seriously.

The political economy of the media shows similarities with other sectors of the Latin American economy in that there is a high concentration of media ownership in the hands of the few. This concentration is being challenged by the proliferation of grassroots, community, indigenous and bottom-up media initiatives. These media spaces are proving crucial in enabling ordinary people to express their views publicly in ways that make a difference. At the same time, corporate media are used in a range of ways by consumers and are often put to alternative or oppositional uses, so it is important to avoid a binary distinction between mainstream and alternative media.

Media and communication technologies have much empowering potential for communities and individuals. They are not to be understood deterministically, however, as important limitations and contradictions often emerge. Media consumption and production are embedded in both structures of power and the cultures of everyday life, and close attention to the geographical specificities of media practice and its relational outcomes in concrete material sites is therefore essential.

Summary

- Media, communication and popular culture are central to the ways in which Latin Americans make development.

- Latin America exports cultural products such as telenovelas as well as commodities.

- The cultural imperialism thesis is rejected by scholars in cultural studies for its failure to come to terms with the complexities of media flows in conditions of globalization.

- Many Latin Americans want to make their own media, and as a result community and grassroots media are flourishing.

- The Internet is changing the way struggles for development and social justice are being fought. This change was clearly demonstrated in the Zapatista rebellion, which enjoys a massive Internet presence.

Discussion questions

1. Outline what is meant by cultural imperialism and why it is a flawed thesis for understanding the Latin American mediascape.

2. Why is access to media and communication technologies a development issue?

3. What are the benefits of community media for marginalized populations?

4. What is the relationship between telenovelas and Latin American development?

5. In your opinion, does Yoani Sánchez deserve the global acclaim she has received?

Further reading

del Sarto, A., Ríos, A. and Trigo, A. (eds) (2004) *The Latin American Cultural Studies Reader*. Durham, NC: Duke University Press. A large collection of some of the most important scholars in Latin American cultural studies, covers historical developments in the field and key theoretical debates.

Fox, E. and Waisbord, S. (eds) (2002) *Latin Politics, Global Media*. Austin, TX: University of Texas Press. An edited collection that explores the transformation of Latin American media in conditions of globalization.

García Canclini, N. (2001) *Consumers and Citizens: Globalization and Multicultural Conflicts*. Minneapolis, MN: University of Minnesota Press. Written by one of the leading theorists in Latin American cultural studies, this book outlines the convergence between cultural consumption and citizenship.

Martín-Barbero, J. (1993) *Communication, Culture and Hegemony: From the Media to Mediations*. London: Sage. Another prominent name in Latin American cultural studies, this text provides a ground-breaking analysis of the role of media audiences, popular culture and everyday life.

Rowe, W. and Shelling, V. (1991) *Memory and Modernity: Popular Culture in Latin America*. London: Verso. This text foreshadowed many of the theoretical developments to come in Latin American cultural studies surrounding modernity, tradition, globalization and hybridity. An engaging and sophisticated text.

Shaw, L. and Dennison, S. (eds) (2005) *Pop Culture Latin America! Media, Arts, and Lifestyle*. Santa Barbara, CA: ABC-CLIO. An excellent reference book containing short essays on Latin American popular culture, including musical styles, festivals, visual arts, sport, literature, cinema and television.

Stavans, I. (ed.) (2010) *Telenovelas*. Santa Barbara, CA: Greenwood. An edited collection that introduces the reader to the diversity and cultural significance of the Latin American telenovela.

Films

A Place Called Chiapas (1998), directed by Nettie Wild (Canada). A Canadian documentary on the Zapatista rebellion, which includes an on-camera interview with Subcomandante Marcos.

Chavez: Inside the Coup (2003), directed by Kim Bartley and Donnacha Ó Briain (Ireland). A documentary about the attempted coup against Hugo Chávez in Venezuela in 2002 and the role played by the private media.

Favela Rising (2005), directed by Jeff Zimbalist and Matt Mochary (USA). A documentary set in the favelas of Rio de Janeiro that looks at the use of popular culture, especially hip hop and dance, to challenge the violence of everyday life.

Un Poquito de Tanta Verdad/A Little Bit of So Much Truth (2007), directed by Jill Friedberg (USA). Based on the popular uprisings that took place in Oaxaca, Mexico in 2006, this film explores how ordinary citizens became savvy media producers to denounce corruption and governmental social neglect.

Websites

www.agoratv/org
The website of Agora TV – a community television production collective in Argentina that aims to cover issues ignored by mainstream media and to challenge the concentration of media ownership that occurred during the period of neoliberalization. It broadcasts over the Internet.

www.zonalatina.com
All kinds of links to information about Latin American media, including television, newspapers, music as well as articles about the media.

www.eco.utexas.edu/faculty/Cleaver/zapsincyber.html
A website maintained by Harry Cleaver at the University of Texas – contains links to Zapatista resources available on the Internet.

www.internetworldstats.com/stats.htm
Internet World Stats, a website that has reliable and regularly updated statistics on Internet usage around the world, broken down by region and country.

9 Decolonizing Latin American development

Learning outcomes

By the end of this chapter, the reader should:

- Be able to evaluate Latin America's contemporary situation and the extent to which development challenges are being overcome
- Be able to review Latin America's relations with other parts of the world, especially with the United States and China
- Be able to discuss the decolonial option as a way of thinking through and beyond development

Changing times in Latin America

In the 1980s and 1990s, the international media coverage of Latin American development painted a very negative picture of a continent mired in growing poverty and inequality, intractable civil wars, massive social unrest and escalating external debt. Of course, as this book has emphasized, everyday life in the region was far more complex and diverse and even during the hardest decades there were many positive ways in which low-income Latin Americans found dignity and happiness. But for many outside the region, an impression of hopelessness often prevailed.

In the second decade of the twenty-first century, an alternative picture of the region is gaining momentum. It is now the US and parts of Europe that are struggling with unpayable external debts, recession, drastic public sector cuts and budget deficits. It is now Greeks and Spaniards rather than Venezuelans and Bolivians that are taking to the streets and mobilizing against austerity. While poverty and inequality remain persistent for many Latin Americans and in some areas urban insecurity and violence continue to tear the region's social fabric, there are also, as recent global media coverage illustrates, significant sources of optimism.

Latin America is providing the world with inspiring new examples of how to do development, in areas such as constitutional reform, multilateral governance and urban sustainability, in many cases led by indigenous populations who have been marginalized for centuries. For example, a 2012 article in British newspaper *The Guardian* asks whether Ecuador could be "the most radical and exciting place on earth" (Ghosh 2012). It describes how the Ecuadorian government is promoting cultural diversity and human rights, protecting the environment, standing up to multinational oil companies and spending on health, education and housing.

While the US was struggling with a drastic and politically polarizing budget deficit of some $15 trillion, the Bolivian economy was posting a budget surplus. Some commentators believe that Argentina's recovery and return to growth after defaulting on debt and investing in social programmes should provide the Greeks with a viable model for addressing their own debt-fuelled economic collapse (Valente 2011). The Brazilian economy has been growing rapidly for a number of years. In 2011, as Portugal struggled, it looked as if Brazil would have to come to the aid of its former colonial master (Wearden 2011) and at the end of 2011, Brazil was asked to *lend* to (rather than borrow from) the IMF, an event that "shows a remarkable shifting of power in the international scenario" (Palhares 2011). Brazil's new leverage enhances the possibility of reform within the IMF and could in time mean the end of conditionality as we know it. In 2012, Brazil overtook the UK as the sixth largest economy in the world.

There is some evidence that south–south cooperation is beginning to displace and challenge north–south cooperation. This includes developments within Latin America, such as the creation of ALBA, which has enabled mutually beneficial intra-continental exchange and cooperation to flourish. New extra-continental alliances that do not include the US or Europe are also growing. Brazil, South Africa and India now make up a G3 with rates of economic growth and social investment that are starting to put the G8 countries to shame. US and European investment continues to be significant in the region but is challenged by substantial investment from China and other Asian countries. As a result of economic crisis at home, in 2012 Spain cut its aid budget to Latin America by more than 70 per cent (AFP 2012), a move that substantially curtails Spain's role in making development in its former colonies.

In the nineteenth century, Latin America was being pushed to the south by a hegemonic United States. Today, as the Latin American population resident in the United States grows, Latin America is once again moving northwards. This process has important implications for Latin American development. There are clearly important geopolitical, geoeconomic and cultural shifts under way that are reconfiguring Latin America's place in the world.

This chapter will outline the key contemporary features of Latin American development, in particular how global and continental transformations are reshaping development as we understand it. After a discussion of some of the salient dimensions of contemporary Latin American development, including changing relations with the US and China, it will consider whether we might be moving from development to decoloniality.

Globalization and development in Latin America in the twenty-first century

As this book has made clear, globalization is both something that shapes Latin American development and something shaped by Latin American development, and these processes do not unfold in a predictable or predetermined way. While there are many global forces of an economic, political and cultural nature that complicate Latin America's attempt to shake off the legacies of colonialism and continue to restrict the ability of Latin American governments and citizens to put their countries or communities on a more sustainable trajectory, globalization is not only a top-down externally imposed phenomenon. It is being made as much from the bottom up, by indigenous peoples, by *campesinos*, by small entrepreneurs, by men and women in low-income neighbourhoods, by environmental activists, by elected politicians and by media makers, as it is from the top down by multinational companies or multilateral organizations such as the IMF (Flusty 2004).

In economic terms, there is plenty of evidence that capital, especially multinational capital, continues to concentrate wealth in alarming ways. For example, successful Nicaraguan instant coffee company Café Presto, which provided serious local competition to Nescafé, was simply purchased by European food giant Nestlé, the makers of Nescafé, which now enjoys and expatriates the profits of an imagineered brand differentiation. Similarly, the Mexican beer company FEMSA, which began production in 1890 as the Cuauhtémoc brewery and makes popular Mexican beers such as Dos Equis and Tecate, was recently purchased by Heineken. But, as Dangl (2010) notes, such processes do not go unchallenged and are responded to in a range of complex and creative ways, and he points to the parallel existence of countervailing tendencies. These include the decision by Bolivian president Evo Morales to launch a new drink onto the Bolivian market called Coca Colla, containing legally grown coca.

In the popular eateries or *fritangas* of Managua, the locals are just as likely to order traditional drinks such as *pinolillo* made of corn, cacao and cinnamon as they are to order a Coca-Cola. While many Central Americans frequently patronize Pizza Hut and McDonald's, this phenomenon has not resulted in the disappearance of traditional foods and traditional food sellers from the urban landscape (see Figure 9.1).

Changing economic and political conditions both within and beyond Latin America are also providing Latin American governments with new sources of leverage and political assertiveness and underscore the multidirectionality of globalizing processes. In 2008, Rafael Correa terminated the lease on a military base in Ecuador held by the US, stating that the US would be welcome to maintain a military base in Ecuador as long as Ecuador could open a military base in Miami. In 2008, President Evo Morales threw the US Drug Enforcement Agency (DEA) out of Bolivia and then, in 2011, he withdrew his country from the 1961 Vienna Convention on Narcotic Drugs, because it contravened the Bolivian Constitution, which ends a ban on coca chewing and the rights of indigenous peoples (Cote-Muñoz 2011).

In some places, those accused of human rights abuses are slowly being brought to justice. In 2012, after years of campaigning by Guatemalan Mayan groups in solidarity with US activists, General Ríos Montt was finally made to face trial on charges of genocide. At the same time, Colombia continues to collaborate militarily with the United States. In 2009, Colombia signed an agreement which gave US military personnel access to seven Colombian military bases, a move that provoked tensions in the region and was later declared unconstitutional (Associated Press 2010). This declaration does not alter the US military presence in Colombia, however, as this presence is covered by earlier agreements.

Viewed in broader geopolitical terms, the economic, political and cultural landscape is however in a state of flux, particularly given changing relationships between Latin America and other parts of the world. The United States continues to be a significant foreign policy actor in Latin America but one that is challenged in the region by intra-continental trade and cooperation among Latin American countries and by Latin American relations with countries hostile to the US, such as Iran, as well by the rapidly developing trading and diplomatic relations between China and

Figure 9.1 *In Quetzaltenango, Guatemala, one can find both McDonald's and traditional foods. McDonald's in Quezaltenango (top) and traditional food stalls in Quezaltenango (bottom)*

Source: Marney Brosnan.

Latin America. While relations between Latin America and Europe continue to be important, changing relations with the United States and developing relations with China have the most far-reaching implications for the region.

Latin America and the United States

For a long time, one of the key development problematics facing Latin America was its complex relationship with its northern neighbour. As we have seen, economic and political intervention in Latin America by the United States has often had tragic outcomes. The CIA has worked to overthrow politicians that threaten US interests, US banana companies have contributed to the destruction of biodiversity and the pollution of lands and waterways, US military aid has been used to torture and disappear Latin Americans deemed subversive. But now it appears that the relationship is beginning to change, for a number of reasons.

The US is no longer the economic powerhouse it once was, and is being challenged economically by Latin American economic growth, especially in countries such as Brazil, and by the "pink tide" policies more generally which are undermining the ability of the United States to use the IMF and the World Bank to forge and protect its own interests in the region. As we have seen, US oil interests in the region are being lost as Latin American governments opt for nationalization to facilitate investment in social programmes. It now appears likely that there are substantial oil reserves off the coast of Cuba, and drilling for oil is under way. As Sandels (2011: 40) writes, if estimates of oil reserves prove to be correct, Cuba's economic status would be dramatically transformed and it "would send the US sanctions policy into the dustbin of imperial miscalculations".

Despite or because of this changing economic landscape, the US still finds ways to intervene in Latin America. Many hoped that the election of Barack Obama in 2008 would initiate a new era for US–Latin American relations. Longstanding aspects of US foreign policy towards Latin America, such as the US policy towards Cuba, are looking increasingly anachronistic and are gradually being phased out, although not nearly quickly enough in the eyes of many, and for many scholars there is more continuity than change (see Weisbrot 2011).

As discussed in Chapter 4, one of the main ways in which US military intervention in Latin America continues is through the "war on drugs". This policy provides the US with a pretext to maintain its military presence in Colombia. The focus on reducing supply in Latin America rather than demand in the United States through militarization and crop eradication is now increasingly viewed as a failure, as it has harmed poor subsistence farmers while traffickers continue to profit and addiction rates remain high. The "war on drugs" is fuelling drug-related violence and instability without tackling the problem, and is increasingly viewed as a war that cannot be won. In the meantime, it is now estimated that there is a drug trade related killing on average every hour in Mexico (Tuckman 2011) and crackdowns in Mexico merely encourage traffickers to shift operations elsewhere in the region, recruiting gang members. Consequently, drug-fuelled violence and militarization have now taken hold in Honduras. In the Mosquitia region of Honduras, in May 2012, the DEA shot dead four innocent people in a counter-narcotics raid (Sackur 2012). The voices calling for an alternative approach to drug trafficking including decriminalization are therefore growing.

Recent releases of WikiLeaks cables have revealed a number of continuities in US–Latin America relations that are underpinned by imperialist relations of power and that hinder rather than enable development. As Soltis (2011b) reports, one cable revealed how the US attempted to prevent post-quake Haiti joining Petrocaribe, a regional initiative that would have given Haiti access to subsidized Venezuelan oil. It would have saved the Haitian government $100 million that could have been invested in health care and education. In order to protect US textile companies in Haiti, the US also intervened to keep the minimum wage at 31 cents an hour.

There is evidence that US multinationals continue to attempt to protect their interests in anti-developmental and neocolonial ways. In February 2012, residents from the community of Salado Lislis in Honduras alleged that Dole (Standard Fruit) had sent in the army to evict them from their land. According to activists, 200 troops equipped with bulldozers and wearing Dole shirts under bulletproof vests moved into a community, giving residents only 30 minutes to collect their belongings after which their houses and crops were destroyed (La Voz de Los de Abajo 2012).

Overall US political and military intervention in Latin America has decreased, not only because of growing awareness of the harm it has done, but also because the US has been preoccupied with fighting wars in other parts of the world, particularly Iraq and Afghanistan. The "war on terror" does however affect US–Latin American relations. In order to further leverage their political and economic independence from the United States, the "pink tide" governments are developing diplomatic relations with countries such as Iran. Venezuelan president Hugo Chávez and Nicaraguan president Daniel Ortega flaunt their controversial friendship and economic and political ties with Iranian leader Mahmoud Ahmadinejad in front of the global media, which clearly provokes the US administration, as is the intention. But it also provides the United States with the discursive justification for continued intervention in the region, and the US has attempted to link the "pink tide" to its fight against terrorism in the Middle East.

In July 2011, the US Congress Subcommittee on Counterterrorism and Intelligence expressed concern about what it referred to as Hezbollah in Latin America, noting the penetration of Iran in Latin America fuelled by the close relationship between Venezuela and Iran. The participants, who included Roger Noriega, co-author of the Helms-Burton law which strengthened the US trade embargo imposed on Cuba, expressed concern for example that it was possible to fly from Caracas to Teheran with only one stopover. In his testimony, Noriega linked the Islamic influence on the Americas to drug trafficking.[1]

Latinos in the United States

One interesting phenomenon that shapes US–Latin American relations in complex ways is the growth and increased visibility of the Latino population in the United States. The 2010 US census showed that the Latino or Hispanic population in the US is growing faster than any other demographic and is now the largest ethnic minority. There are currently more than 50 million people of Latin American descent living in the US, up from 35 million in 2000. While the majority (30 million) are Mexicans, immigration from other Latin American countries is growing more rapidly. The Latino population is concentrated in Texas, California and Florida but is growing in every single US state (see Ennis *et al.* 2011).

Such dramatic demographic changes and the presence of Latinos in all walks of life are having significant implications for both the United States and the countries of Latin America. While there is plenty of evidence that Latin Americans are well integrated into US culture, Latin American culture, including language, food, sports teams, cultural activities and religious celebrations, is now ubiquitous in the US, especially in cities such as Los Angeles and Miami. There are many prominent Latinos playing leadership roles in sports, culture and politics. Barack Obama has appointed two Latinas of working-class origin to high profile political roles: Mexican American Hilda L. Solís as Secretary of Labor and Puerto Rican descended Sonia Sotomayor as a Supreme Court judge (see Figure 9.2). Other prominent Latinos in the US include Antonio Villaraigosa, Mexican American mayor of Los Angeles from 2005 to 2013; Cuban American fashion designer Narciso Rodríguez; NFL stars Victor Cruz and Aaron Hernández, who played on opposing teams in the 2012 Super Bowl; and Costa Rican astronaut Franklin Ramón Chang-Díaz.

Figure 9.2 US President Barack Obama has appointed two women of Latin American descent to high profile political roles: Hilda L. Solís, Mexican American Secretary of Labor (left); Sonia Sotomayor, Puerto Rican-descended Supreme Court Judge (right)

Sources: Department of Labor, Wikimedia Commons (released into public domain) (left-hand picture); Steve Petteway, Collection of the Supreme Court of the United States, Wikimedia Commons (released into public domain) (right-hand picture).

There are of course many more, and all over the US Latin Americans are appearing in the media, running businesses, joining the US military and teaching in schools and universities. Millions more, some of whom have no legal status, are making massive and often invisibilized contributions to the US economy and everyday social reproduction by cleaning homes, caring for children, tending gardens, working in restaurants and factories and picking fruit and vegetables.

The massive presence of Latinos in the United States is transforming the formal political sphere and US culture more broadly. It is clear for example that media companies can no longer afford to ignore Latino viewers and US politicians can no longer afford to ignore Latino voters. A February 2012 issue of *Time* magazine had on its cover "Yo decido: Why Latinos will pick the next president". There is no doubt that voters of Latin American descent were a decisive demographic in the re-election of Barack Obama in 2012 (Foley 2012). After centuries of coups and military interventions in Latin America by the US to remove presidents and governments that the US administration did not like, there is a certain irony to the idea that the outcome of US elections is now partially determined by Latin Americans.

These demographic changes are not however warmly embraced by all US Americans, and many US-based Latinos continue to be subject to discrimination and marginalization. Anxieties within the US about changing demographics and economic recession have unleashed a fierce anti-immigrant sentiment in some parts of the country, which is especially targeted at Latin Americans. In 2010 and 2011, Arizona, Georgia, Alabama and South Carolina all passed laws targeting illegal immigrants. Fear of being targeted in the US states with draconian anti-immigration laws led many illegal immigrants to leave their jobs. The result was a shortage of labour in states such as Georgia, and consequently crops and fruit rotted in the ground and on the trees.

There have been attempts to introduce legislation that would grant residency and citizenship to a number of illegal residents but these continue to meet with opposition from politicians. For example, the DREAM (Development Relief and Education for Alien Minors) Act was introduced in order to provide residency and citizenship for Latin Americans who were brought to the US illegally as minors by their parents and who fulfilled certain college or military service requirements. Interestingly, the bill was supported by military leaders, who were facing a shortage of enlistees, but has so far failed to pass as opponents argue it would encourage further illegal immigration.

Whatever one's views on illegal immigration, it is clear that the vast majority of Latin Americans migrate to the United States in search of employment and a better life and that many of them end up doing the jobs that many US Americans do not wish to do, working as nannies, gardeners and domestic servants, as employees in fast food restaurants and factories and as agricultural labourers and construction workers. They have also been driven to migrate as a result of US-led policies that have exacerbated hardship and misery such as structural adjustment, military aid and the escalation of the drugs war. In addition to their decisive contribution to the US economy, they send almost US$60 billion worth of remittances back to their families in Latin America, which, as Cohen (2011) observes, enables Latin Americans to improve their access to health care, send their children to school and invest in homes, infrastructure and local businesses. In other words, remittances go some way towards compensating for the failures of development and allow some family members to remain in their countries of origin.

In recent years, illegal immigration from Latin America to the United States has slowed and has now reached its lowest levels since the 1950s. According to Cave (2011), a number of factors account for this decline. They include the fact that legal immigration has become easier and crossing illegally has become more dangerous and more expensive as a result of the militarization of the border area and the activities of drug traffickers. Improved educational and employment opportunities at home and recession in the US are also important.

China and Latin America

One of the key global reconfigurations of geoeconomic power in the world with substantial implications for Latin America has been the economic rise of China. China's position in the world began to shift substantially after the trade liberalization measures implemented by the government of Deng Xiaoping in the late 1970s.

Relations between China and Latin America are not new. Indeed, Latin America was targeted by China's ideological agenda and the continent had its own variant of Maoism, embodied most clearly in the Shining Path of Peru. But as Fernández Jilberto and Hogenboom (2010a) point out, their nature has changed substantially and today relations are mainly economic rather than ideological. Latin America was central to the negotiations that preceded China's entry into the WTO in 2001, and trade between the two regions has grown exponentially. Countries such as Brazil supported China's entry in the WTO because they saw it as creating a more united opposition to US and EU agricultural subsidies, which were putting Latin America's agriculture at a disadvantage.

China's economic rise is both positive and negative for Latin America. Cheap textiles and manufactured goods produced in China have flooded Latin American markets and have been negative for export processing production and for small producers. On the other hand, China's huge demand for energy, minerals and agricultural products has kept the prices of commodities high and has enabled Latin America to find new markets. China is importing large quantities of oil, copper, steel and soy from Latin America, mainly from Brazil, Argentina, Chile and Peru. It is also providing some foreign direct investment (FDI) in these sectors. High copper prices fuelled by demand by China have for example enabled Chile to maintain a trade surplus and invest in social programmes and higher education, including large numbers of government scholarships to allow Chileans to gain Masters and PhD degrees overseas. Trade with China, along with sound domestic policies, has been a key factor in the dramatic growth of the Brazilian economy. Since 2000, Brazil's exports to China, mostly soy and iron ore, have increased by US$28.8 billion, outstripping imports from China by more than US$5 billion (Sami 2011).

In the past decade there have been dozens of bilateral agreements between China and Latin American countries, fomenting economic, scientific and technological cooperation.

Some Latin American countries, particularly Mexico and those of Central America, have welcomed China less warmly than many countries in South America. Nicaragua for example has chosen to develop relations instead with Taiwan, which has been an important source of aid and investment to Nicaragua, which meant it had to formally end its diplomatic relations with China (Fernández Jilberto and Hogenboom 2010a; Aguilera Peralta 2010).

Even in South America, where trade is well established and growing dramatically, there is ongoing concern about the potential negative impacts of China on local economies. Brazil for example has strengthened its levels of protectionism in response to fears that China is engaging in the dumping of goods below the cost of production in order to put Brazilian competition out of business (Sami 2011). It is apparent that China's interest in Latin America is driven by Chinese rather than mutual interests. In particular, China's voracious and growing demand for energy and its serious food security issues at home make Latin America's land and energy production of particular interest. For example, as Lopez-Gamundi and Hanks (2011) report, Chinese food corporation Beidahuang signed an agreement with the provincial government of Río Negro, Argentina, to rent large tracts of land to grow GM crops for export to China. The deal, which provides China with generous tax exemptions, violates the rights of local people and is likely to exacerbate environmental destruction.

There are then a wide variety of "China effects" under way in Latin America, but the significance of new forms of cooperation between China and Latin America is prompting analysts to wonder if the Washington Consensus is being displaced by the Beijing Consensus (Fernández Jilberto and Hogenboom 2010b: 182). It is not at all clear whether contemporary China–Latin America relations are characterized more by decolonization or recolonization.

From development to decoloniality?

One of the key debates within development studies is whether development is merely a continuation of colonialism or imperialism by other means. This book has demonstrated the constraints placed on Latin

America and the harm done to people's lives and livelihoods and to natural environments in the name of development. But we have also seen how development can be harnessed and re-articulated by ordinary people in ways that lead to positive social or political transformation. This book has discussed many of the ways in which this is being achieved, noting in particular the "pink tide" and associated contestations of neoliberalism, the vitality and energy of indigenous and Afro-Latino mobilizations, the geographic displacement of Latin America northwards, the creative harnessing of the new media ecology for democratizing purposes, and new attitudes to environments and natural resources. In this context, it is worth asking whether the idea of Latin American development is becoming less tenable and persuasive. Mignolo (2005) suggests that the Afro and indigenous mobilizations under way are making the idea of *Latin* America obsolete. With respect to development, Escobar (1999: 128, cited in Castro-Gómez 2007: 437) writes that "the idea of development is losing part of its strength. Its incapacity to carry out its promises, together with resistance from many social movements and many communities is weakening its powerful image."

An emerging set of theoretical developments within Latin American cultural studies is a potentially fruitful way to rethink the concept of Latin American development and its multiple mutations. These developments have coalesced into what is referred to as the de-colonial option or the modernity/coloniality/decoloniality (MCD) research paradigm. It is a body of literature being advanced by scholars such as Walter Mignolo, Arturo Escobar, Aníbal Quijano, Catherine Walsh and Nelson Maldonado-Torres. While this paradigm shares some theoretical sympathy with theoretical developments in postcolonialism, it draws primarily on bodies of Latin American indigenous, popular and subaltern thought rooted in dependency theory, philosophy of liberation, and debates on Latin American modernity and postmodernity. Its key influences include Carlos Mariátegui, Enrique Dussel, Rodolfo Kusch and Gloria Anzaldúa. It is also inspired by what Escobar (2007) refers to as "landmark experiences of decolonization" such as the 1780 Tupac Amaru rebellion in Peru and the 1804 Haitian slave rebellion, and by contemporary decolonial social movements in Latin America which include the Zapatista movement in Chiapas, the leadership of Evo Morales in Bolivia, and a variety of indigenous, Afro-Latin and African American social movements that have been asserting themselves throughout the Americas. The key idea is that modernity and coloniality are inextricably linked – that modernity began not with the Enlightenment but with the conquest of America in the fifteenth century.

Given that being modern in Latin America meant dominating others, there is then "no modernity without coloniality" (Escobar 2007: 185).

Even after independence, coloniality became embedded in Latin American societies because of the ways in which creole elites marginalized and dominated the Indian and African populations (Mignolo 2005). The failure to recognize coloniality as constitutive of modernity inhibits understanding and so the decolonial option attempts to promote a different kind of thinking, called variously border thinking, border epistemology or interculturality (see Mignolo 2007a, 2007b; Escobar 2007; Quijano 2007; Walsh 2007). This is a non-linear and spatial kind of thinking, which embraces other ways of knowing and being and reveals the "possibility of thinking from different spaces which finally breaks away from eurocentrism as sole epistemological perspective" (Escobar 2007: 188).

This is not a dismissal of modernity, but a transmodernity or thinking about modernity from the perspective of the excluded other (see also Dussell 1976). An example of this is the embrace of the Aymara concept of *buen vivir* which means to live well and is based on collective reciprocal relationships and solidarities. It is therefore distinct from a capitalist and individualistic notion of living better. It produces different and more inclusive ways of organizing economies and relating to nature and community, especially in the Andes where the concept originates. Walsh (2010) suggests that while *buen vivir* has some similarities with western understandings of sustainable development and is therefore being mobilized in a more hybrid way, the inclusion of the concept in the new Ecuadorian Constitution stimulates "an 'interculturalizing' unprecedented in the country as well as the Latin-American region. It requires the general populace to think and act 'with' ancestral principles, knowledges, and communities" (p. 19).

This body of theory is being applied to a range of decolonial projects in Latin America, from black/Afro-Latin resistance movements to indigenous media and intercultural arts projects, solidarity economies, constitutional reform and resource extraction. It is useful because it does not deny the hardships and suffering that come with coloniality and capitalism, but also finds spaces of hope and the moments in which an-other thinking is breaking through, being heard, gaining momentum.

Keeping a sense of the ongoing injustices and finding sources of optimism about the possibilities for decolonization in tension is important. As noted above, there are examples of both the intensification and the unravelling of neoliberalism in Latin America and the continent, in García Canclini's (2002: 43) words, is full of movements such as the

Zapatistas and the MST "that have neither triumphed or failed". While there are many positive dimensions to Latin American development, many Latin Americans are still struggling with hunger, poverty, urban insecurity and drug-related conflict, so working in Latin American development – being part of the struggle to attend to the factors that induce suffering or deprive people of their dignity – continues to be important. We therefore need to identify setbacks to the decolonial project as well as empowering processes that can be built on and that are likely to proliferate across the region.

This is an exciting time to be studying Latin American development. Contemporary developments in the continent are disrupting the fundamental idea of development as it was conceived in the postwar period. Increasingly, at the present moment, Latin America is showing the so-called first world that it has particular expertise to offer on how to create a better world. From the concept of *buen vivir* in constitutional governance to investing in rather than cutting public services, to participatory budgeting, to indigenizing the formal political sphere, to putting natural resources to work for rather than against people and environments, Latin America abounds with creative examples of how to do development differently. It seems that first world development practitioners and experts have much to learn from Latin America. So engagement with Latin American development through scholarship, practice, policy or activism should be based on solidarity and mutual respect and exchange rather than on a one-way delivery of expertise or technical knowledge.

One of the key challenges for students, practitioners and scholars of Latin American development at the current time is to think very carefully about how to practise border thinking in our work and to consider how we might approach the question of Latin American development from the perspective of the excluded other. If we can do that, development may start to look very different and its obsolescence will be accelerated.

Summary

- The international media coverage of Latin America is very different today compared with the 1980s and 1990s.

- US intervention in Latin America has decreased but the US still intervenes in the name of the "war on drugs" and the "war on terror".

- The Latin American population in the United States has increased dramatically, with important implications for Latin American development.

- Relations between Latin America and China are growing and have both positive and negative outcomes for the region.

- Latin America is providing the world with inspiring new ways to do development.

- A body of literature known as the decolonial option urges a rethinking of contemporary development concepts.

Discussion questions

1. What does it mean to say that globalization processes in Latin America are multidirectional?
2. How is Latin America's relationship with the United States changing?
3. What effect does the growing presence of Latino culture in the United States have on Latin American development?
4. Why are some Latin Americans concerned about Chinese investment in the region?
5. Is the idea of Latin American development becoming obsolete?

Note

1 His full testimony can be found here: http://homeland.house.gov/sites/homeland.house.gov/files/Testimony%20Noriega.pdf

Further reading

Ennis, S. R., Ríos-Vargas, M. and Albert, N. G. (2011) *The Hispanic Population: 2010*. Washington, DC: United States Census Bureau. A report that analyses what the 2010 US census has revealed about the Latino population in the United States.

Latin American Perspectives 38(4) (2011). A special issue of this journal which explores what the election of Barack Obama in the United States means for Latin America and US–Latin American relations.

Fernández Jilberto, A. E. and Hogenboom, B. (eds) (2010) *Latin America Facing China: South–South Relations beyond the Washington Consensus*. New York:

Berghahn Books. An edited collection that explores various facets of China Latin American relations and how relations with China differ from country to country.

Mignolo, W. and Escobar, A. (eds) (2010) *Globalization and the Decolonial Option*. London: Routledge. An edited collection that provides an invaluable introduction to the decolonial option with chapters from the main intellectual contributors to this field. It is also available as a special issue of the journal *Cultural Studies* published in 2007.

Website

http://mycuentame.org
Described as the official page of the ¡Latino Instigators!, Cuéntame uses Facebook and the Internet more broadly to discuss issues affecting Latinos. It denounces injustice in the application of US immigration legislation and promotes Latino activism and artistic endeavour.

 # References

Achtenberg, E. (2012) Bolivia: TIPNIS communities divided as road consultation begins. *NACLA* 5 August. https://nacla.org/blog/2012/8/5/bolivia-tipnis-communities-divided-road-consultation-begins (Accessed 20 August 2012).

AFP (2012) Recorte español afectará la ayuda a Latinoamérica. *La Razón* 1 April. http://www.la-razon.com/economia/Recorte-espanol-afectara-ayuda-Latino america_0_1588041185.html (Accessed 20 August 2012).

Aguilera Peralta, G. (2010) Central America between two dragons: Relations with the two Chinas. In A. E. Fernández Jilberto and B. Hogenboom (eds) *Latin America Facing China: South–South Relations beyond the Washington Consensus.* New York: Berghahn Books, pp. 167–180.

Akass, K. and McCabe, J. (2007) Not so ugly: Local production, global franchise, discursive femininities, and the Ugly Betty phenomenon. *Flow TV* 5(7). http://flow tv.org/2007/01/not-so-ugly-local-production-global-franchise-discursive-femininities-and-the-ugly-betty-phenomenon (Accessed 12 January 2012).

Albó, X. (1991) El retorno del indio. *Revista Andina* 9(2): 299–345.

Albó, X. (1996) Making the leap from local mobilization to national politics. *NACLA Report on the Americas* 29(5): 15–20.

Amazon Watch (2010) Ecuador signs historic deal to keep oil in the soil and CO_2 out of the atmosphere. *Upside Down World* 3 August. http://upsidedownworld.org/main/news-briefs-archives-68/2623-ecuador-signs-historic-deal-to-keep-oil-in-the-soil-and-co2-out-of-the-atmosphere (Accessed 9 November 2011).

Anaya, S. J. and Grossman, C. (2002) The case of Awas Tingni v. Nicaragua: A new step in the international law of indigenous peoples. *Arizona Journal of International and Comparative Law* 19(1): 1–15.

Anderson, M. (2009) *Black and Indigenous: Garifuna Activism and Consumer Culture in Honduras.* Minneapolis, MN: University of Minnesota Press.

Andolina, R., Laurie, N. and Radcliffe, S. (2009) *Indigenous Development in the Andes: Culture, Power, and Transnationalism.* Durham, NC: Duke University Press.

André, R. (2011) The invisible war against Afro-Colombians. *Americas Quarterly Blog* 16 March. www.americasquarterly.org/node/2322 (Accessed 12 December 2011).

Angel, A. and Macintosh, F. (1987) *The Tiger's Milk: Women of Nicaragua.* London: Virago.

Appadurai, A. (1996) *Modernity at Large: Cultural Dimensions of Globalization*. Minneapolis, MN: University of Minnesota Press.

Appelbaum, N. P., Macpherson, A. S. and Rosemblatt, K. A. (2003) Introduction: Racial nations. In N. P Appelbaum, A. S. Macpherson and K. A. Rosemblatt (eds) *Race and Nation in Modern Latin America*. Chapel Hill, NC: University of North Carolina Press, pp. 1–31.

Archetti, E. P. (1999) *Masculinities: Football, Polo and the Tango in Argentina*. Oxford: Berg.

Arizpe, L. and Aranda, J. (1986) Women workers in the strawberry agribusiness in Mexico. In E. Leacock and H. I. Safa (eds) *Women's Work: Development and the Division of Labour by Gender*. New York: Bergin and Garvey, pp. 174–193.

Asher, K. (2009) *Black and Green: Afro-Colombians, Development, and Nature in the Pacific*. Durham, NC: Duke University Press.

Assies, W. (2003) David versus Goliath in Cochabamba: Water rights, neoliberalism, and the revival of social protest in Bolivia. *Latin American Perspectives* 30(3): 14–36.

Associated Press (2010) Colombian agreement over US military bases "unconstitutional". *The Guardian* 18 August. www.guardian.co.uk/world/2010/aug/18/colombia-us-bases-unconstitutional (Accessed 23 August 2012).

Augier, A. and Bernstein, J. M. (1951) The Cuban poetry of Nicolás Guillén. *Phylon* 12(1): 29–36.

Babb, F. E. (2009) Neither in the closet nor on the balcony: Private lives and public activism in Nicaragua. In E. Lewin and W. L. Leap (eds) *Out in Public: Reinventing Gay/Lesbian Anthropology in a Globalizing World*. Malden, MA: Wiley-Blackwell, pp. 240–255.

Bakewell, P. (2011) Colonial Latin America. In J. K. Black (ed.) *Latin America: Its Problems and Its Promise: A Multidisciplinary Introduction*. Boulder, CO: Westview Press, pp. 77–85.

Ballvé, M. (2004) The battle for Latino media. *NACLA Report on the Americas* 37(4): 20.

Barker, C. (1997) *Global Television: An Introduction*. Oxford: Wiley-Blackwell.

Barker, M. (1989) *Comics: Ideology, Power and the Critics*. Manchester, UK: Manchester University Press.

Barndt, D. (2011) *¡VIVA! Community Arts and Popular Education in the Americas*. New York: SUNY Press.

Barrig, M. (1989) The difficult equilibrium between bread and roses: Women's organizations and the transition from dictatorship to democracy in Peru. In J. S. Jaquette (ed.) *The Women's Movement in Latin America: Feminism and the Transition to Democracy*. Boston: Unwin Hyman, pp. 114–148.

Barrionuevo, A. (2008) Left-leaning president's election gives hope to landless Paraguayans. *New York Times* 13 October. www.nytimes.com/2008/10/14/world/americas/14paraguay.html (Accessed 6 November 2008).

Barton, J. R. (1997) *A Political Geography of Latin America*. London: Routledge.

Bebbington, A. (2009) Latin America: Contesting extraction, producing geographies. *Singapore Journal of Tropical Geography* 30: 7–12.

Bebbington, A., Hinojosa, L., Bebbington, D. H., Burneo, M. L. and Warnaars, X. (2008) Contention and ambiguity: Mining and the possibilities of development. *Development and Change* 39: 887–914.

Bebbington, A. and Williams, M. (2008) Water and mining conflicts in Peru. *Mountain Research and Development* 28(3–4): 190–195.

Beck, J. (2006) The rebirth of Bolivia in a constituent assembly: Is this what democracy looks like? *Upside Down World* 8 August. http://upsidedownworld.org/main/bolivia-archives-31/384-the-rebirth-of-bolivia-in-a-constituent-assembly-is-this-what-democracy-looks-like (Accessed 20 December 2011).

Beder, S. (2002) BP: Beyond Petroleum? In E. Lubbers (ed.) *Battling Big Business: Countering Greenwash, Infiltration and Other Forms of Corporate Bullying.* Totnes, UK: Green Books, pp. 26–32.

Beltran S., L. R. (2004) Communication for development in Latin America: A forty-year appraisal. *Southbound: Development Communicators and Publishers.* www.southbound.com.my/communication/cul-ch.htm (Accessed 11 January 2012).

Bendaña, A. (1999) Nicaragua's structural hurricane. *NACLA Report on the Americas* 33(2): 16–23.

Benería, L. and Feldman, S. (eds) (1992) *Unequal Burden: Economic Crises, Persistent Poverty, and Women's Work.* Boulder, CO: Westview Press.

Benería, L., Floro, M., Grown, C. and MacDonald, M. (2000) Introduction: Globalization and gender. *Feminist Economics.* 6(3): vii–xviii.

Biles, J. J. and Cobos, D. (2004) Natural disasters and their impact in Latin America. In J. P. Stoltman, J. Lidstone and L. M. DeChano (eds) *International Perspectives on Natural Disasters: Occurrence, Mitigation and Consequences.* Dordrecht, The Netherlands: Kluwer Academic Publishers, pp. 281–302.

Bird, A. (2011) Biofuels, mass evictions and violence build on the legacy of the 1978 Panzos massacre in Guatemala. *Upside Down World* 23 March. http://upsidedownworld.org/main/guatemala-archives-33/2965-biofuels-mass-evictions-and-violence-build-on-the-legacy-of-the-1978-panzos-massacre-in-guatemala (Accessed 12 November 2011).

Black, J. K. (2011) Introduction: Latin America leading the learning curve. In J. K. Black (ed.) *Latin America: Its Problems and Its Promise: A Multidisciplinary Introduction.* Boulder, CO: Westview Press, pp. 1–20.

Blandón, M. T. (1994) The impact of the Sandinista defeat on Nicaraguan feminism. In G. Küppers (ed.) *Compañeras: Voices from the Latin American Women's Movement.* London: Latin American Bureau, pp. 97–101.

Boserup, E. (1970) *Women's Role in Economic Development.* London: Allen & Unwin.

Bridge, G. (2004) Mapping the bonanza: Geographies of mining investment in an era of neoliberal reform. *Professional Geographer* 56(3): 406–421.

Brown, G. (2003) To Russia . . . and Bosnia . . . and Latvia, with love: Latin America's quintessential cultural product – the TV melodrama – sees a storyline abroad. *Latin Trade* 11(7): 28–29.

Bryan, J. (2009) Where would we be without them? Knowledge, space and power in indigenous politics. *Futures* 41: 24–32.

Bryan, J. (2010) Force multipliers: Geography, militarism, and the Bowman Expeditions. *Political Geography* 29: 414–416.

Brysk, A. (2000) *From Tribal Village to Global Village: Indian Rights and International Relations in Latin America.* Palo Alto, CA: Stanford University Press.

Buchenau, J. and Johnson, L. L. (eds) (2009) *Aftershocks: Earthquakes and Popular Politics in Latin America.* Albuquerque, NM: University of New Mexico Press.

Budhoo, D. L. (1990) *Enough is Enough: Dear Mr. Camdessus . . . Open Letter of Resignation to the Managing Director of the International Monetary Fund.* New York: New Horizons Press.

Burcher, N. (2012) Facebook usage statistics by country Dec 2008–Dec 2011. www.nickburcher.com/2012/01/facebook-usage-statistics-by-country.html (Accessed 21 January 2012).

Bury, J. (2005) Mining mountains: Neoliberalism, land tenure, livelihoods and the new Peruvian mining industry in Cajamarca. *Environment and Planning A* 37: 221–239.

Calvert, P. (2011) Argentina: Decline and revival. In J. K. Black (ed.) *Latin America: Its Problems and Its Promise: A Multidisciplinary Introduction.* Boulder, CO: Westview Press, pp. 524–537.

Cannon, T. (1994) Vulnerability analysis and the explanation of "natural" disasters. In A. Varley (ed.) *Disasters, Development and Environment.* Chichester, UK: Wiley, pp. 13–30.

Cardoso, E. (1989) Hyperinflation in Latin America. *Challenge* January–February: 11–19.

Carey, M. (2008) The politics of place: Inhabiting and defending glacier hazard zones in Peru's Cordillera Blanca. In B. Orlove, E. Wiegandt and B. H. Luckman (eds) *Darkening Peaks: Glacier Retreat, Science and Society.* Berkeley, CA: University of California Press, pp. 229–240.

Carriere, J. (1991) The political economy of land degradation in Costa Rica. *International Journal of Political Economy* 21(1): 10–31.

Castillo, D. A. (2006) Impossible Indian. *Chasqui: Revista de Literatura Latinoamericana* 35(3): 42–57.

Castro-Gómez, S. (2007) The missing chapter of Empire: Postmodern reorganization of coloniality and post-Fordist capitalism. *Cultural Studies* 21(2–3): 428–448.

Cave, D. (2011) Better lives for Mexicans cut allure of going north. *The New York Times* 6 July. www.nytimes.com/interactive/2011/07/06/world/americas/immigration.html (Accessed 27 September 2011).

CCER (1999) *Propuesta para la reconstrucción y transformación de Nicaragua: Convirtiendo la tragedia del Mitch en una oportunidad para el desarrollo humano y sostenible de Nicaragua.* Managua, Nicaragua: Coordinadora Civil para la Emergencia y la Reconstrucción.

CEH (1999) *Guatemala: Memory of Silence: Report of the Commission for Historical Clarification.* http://shr.aaas.org/guatemala/ceh/report/english/toc.html (Accessed 2 August 2011).

Chacon, J. A. (2011) U.S. intervention in Mexico will make things worse. *The Progressive* 19 August www.progressive.org/mexico_drug_war.html (Accessed 7 August 2012).

Chant, S. (1997a) *Women-Headed Households: Diversity and Dynamics in the Developing World.* Basingstoke, UK: Macmillan Press.

Chant, S. (1997b) Women-headed households: Poorest of the poor? Perspectives from Mexico, Costa Rica and the Philippines. *IDS Bulletin* 28(3): 26–48.

Chuchryk, P. M. (1989) Feminist anti-authoritarian politics: The role of women's organizations in the Chilean transition to democracy. In J. S. Jaquette (ed.) *The Women's Movement in Latin America: Feminism and the Transition to Democracy.* Boston: Unwin Hyman, pp. 149–184.

Clarke, G. (2006) Faith matters: Faith-based organizations, civil society and international development. *Journal of International Development* 18(6): 835–848.

Cohen, T. (2011) One-way ticket or circular flow: Changing stream of remittances to Latin America. *COHA* blog. www.coha.org/one-way-ticket-or-circular-flow-changing-stream-of-remittances-to-latin-america (Accessed 23 August 2011).

Collinson, H. (1990) *Women and Revolution in Nicaragua.* London: Zed Books.

Condit, C. M. (1994) Hegemony in a mass-mediated society: Concordance about reproductive technologies. *Critical Studies in Mass Communication* 11(3): 205–230.

Connell, R. W. (1995) *Masculinities.* Berkeley, CA: University of California Press.

Connolly, P. (1999) Mexico City: Our common future? *Environment and Urbanization* 11(1): 53–78.

Cornia, G. A., Jolly, R. and Stewart, F. (1987) *Adjustment with a Human Face: Country Case Studies.* Oxford: Oxford University Press.

Cornwall, A. (2000) Missing men? Reflections of men, masculinities and gender in GAD. *IDS Bulletin* 31(2): 18–27.

Coronil, F. (2011) The future in question: History and utopia in Latin America. In C. Calhoun and G. Derlugian (eds) *Business as Usual: The Roots of the Global Financial Meltdown.* New York: New York University Press and SSRC, pp. 231–264.

Corrales, J. (2010) Latin American gays: The post-left leftists. *Americas Quarterly* 19 March. http://www.americasquarterly.org/gay-rights-Latin-America (Accessed 12 December 2011).

Corrales, J. and Pecheny, M. (2010) Introduction: The comparative politics of sexuality in Latin America. In J. Corrales and M. Pecheny (eds) *The Politics of Sexuality in Latin America: A Reader on Lesbian, Gay, Bisexual, and Transgender Rights.* Pittsburgh, PA: University of Pittsburgh Press, pp. 1–30.

Cote-Muñoz, N. (2011) Tradition trumps the treaty: Bolivia repeals its ban on coca. *COHA* 12 August. www.coha.org/tradition-trumps-the-treaty-bolivia-repeals-its-ban-on-coca (Accessed 24 August 2011).

Cruz, M. (2010) A living space: The relationship between land and property in the community. *Political Geography* 29: 420–421.

Cuadra, P. A. (1997) *El Nicaragüense.* Managua, Nicaragua: Hispamer.

Cupples, J. (2002) *Disrupting discourses and (re)formulating identities: The politics of single motherhood in postrevolutionary Nicaragua.* Unpublished doctoral thesis, Department of Geography, University of Canterbury, Canterbury, New Zealand.

Cupples, J. (2004) Rural development in El Hatillo, Nicaragua: Gender, neoliberalism and environmental risk. *Singapore Journal of Tropical Geography* 25(3): 343–357.

Cupples, J. (2005) Love and money in an age of neoliberalism: Gender, work, and single motherhood in postrevolutionary Nicaragua. *Environment and Planning A* 37(2): 305–322.

Cupples, J. (2007a) Espacialidades de género y legados revolucionarios en una Nicaragua neoliberal: una interpretación geográfica de la transición de Estado. *Política y Gestión* 10: 125–155.

Cupples, J. (2007b) Gender and Hurricane Mitch: reconstructing subjectivities after disaster. *Disasters* 31(2): 155–175.

Cupples, J. (2009) Rethinking electoral geography: Spaces and practices of democracy in Nicaragua. *Transactions of the Institute of British Geographers* 34(1): 110–124.

Cupples, J. (2011) Shifting networks of power in Nicaragua: Relational materialisms in the consumption of privatized electricity. *Annals of the Association of American Geographers* 101(4): 939–948.

Cupples, J. (2012) Wild globalization: The biopolitics of climate change and global capitalism on Nicaragua's Mosquito Coast. *Antipode* 44(1): 10–30.

Cupples, J. and Glynn, K. (2013) Postdevelopment television? Cultural citizenship and the mediation of Africa in contemporary TV drama. *Annals of the Association of American Geographers* (in press).

Cupples, J. and Larios, I. (2005) Gender, elections, terrorism: The geopolitical enframing of the 2001 Nicaraguan elections. *Political Geography* 24(3): 317–339.

Cupples, J. and Larios, I. (2010) A functional anarchy: Love, patriotism, and resistance to free trade in Costa Rica. *Latin American Perspectives* 37(6): 93–108.

Cupples, J., Glynn, K. and Larios, I. (2007) Hybrid cultures of postdevelopment: The struggle for popular hegemony in rural Nicaragua. *Annals of the Association of American Geographers* 97(4): 786–801.

Dangl, B. (2010) Beer globalization in Latin America: When beer in Mexico is Dutch and chicha in Colombia is popular. *Upside Down World* 19 February. http://upsidedownworld.org/main/mexico-archives-79/2371-globalization-of-beer-in-latin-america-when-mexican-beer-is-dutch-and-chicha-in-colombia-is-popular (Accessed 23 August 2011).

Datta, K. and McIlwaine, C. (2000) Empowered leaders? Perspectives on women heading households in Latin America and Southern Africa. *Gender and Development* 8(3): 40–49.

Davies, J. (2010) Growing fuel instead of food. Agro-fuels in Chiapas. *Upside Down World* 24 August. http://upsidedownworld.org/main/mexico-archives-79/2657-growing-fuel-instead-of-food-agro-fuels-in-chiapas (Accessed 12 November 2011).

de Certeau, M. (1984) *The Practice of Everyday Life*. Trans. S. Rendall. Berkeley, CA: University of California Press.

de la Cadena, M. (2010) Indigenous cosmopolitics in the Andes: Conceptual reflections beyond "politics". *Cultural Anthropology* 25(2): 334–370.

de la Cadena, M. and Starn, O. (2007) Introduction. In O. Starn and M. de la Cadena (eds) *The Indigenous Experience Today*. Oxford: Berg, pp. 1–30.

de Sagahún, B. (1829) *Historia General de las Cosas de Nueva España, Volume 2*. Madrid, Spain: Imprenta del Ciudadano Alejandro Valdés.

de Santana Pinho, P. (2010) *Mama Africa: Reinventing Blackness in Bahia*. Durham, NC: Duke University Press.

de Soto, H. (1989) *The Other Path: The Invisible Revolution in the Third World*. New York, NY: Harper & Row.

del Sarto, A. (2004) Foundations: Introduction. In A. del Sarto, A. Ríos and A. Trigo (eds) *The Latin American Cultural Studies Reader*. Durham, NC: Duke University Press, pp. 153–181.

Delaney, P. L. and Schrader, E. (2000) *Gender and Post-Disaster Reconstruction: The Case of Hurricane Mitch in Honduras and Nicaragua*. Washington, DC: CSPG/LAC Gender Team, The World Bank.

Denevan, W. M. (1992a) Native American populations in 1492: Recent research and a revised hemispheric estimate. In W. M. Denevan (ed.) *The Native Population of the Americas in 1492*, 2nd edn. Madison, WI: University of Wisconsin Press, pp. xvii–xxix.

Denevan, W. M. (1992b) Epilogue. In W. M. Denevan (ed.) *The Native Population of the Americas in 1492*, 2nd edn. Madison, WI: University of Wisconsin Press, pp. 289–294.

Dennis, P. A. (2004) *The Miskitu People of Awastara*. Austin, TX: University of Texas Press.

Dennis, P. A. and Olien, M. D. (1984) Kingship among the Miskito. *American Ethnologist* 11(4): 718–737.

Dominguez, R. (1998) Digital Zapatismo. www.thing.net/~rdom/ecd/DigZap.html (Accessed 12 August 2011).

Dore, E. (1997) The holy family: Imagined households in Latin American history. In E. Dore (ed.) *Gender Politics in Latin America: Debates in Theory and Practice.* New York: Monthly Review Press, pp. 101–117.

Dorfman, A. and Mattelart, A. (1975) *How to Read Donald Duck: Imperialist Ideology in the Disney Comic.* New York: International General.

Drouin, M. (2010) Understanding the 1982 Guatemalan genocide. In M. Esparza, H. R. Huttenbach and D. Feierstein (eds) *State Violence and Genocide in Latin America: The Cold War Years.* London: Routledge, pp. 81–103.

Duffy, G. (2009) "God's gift" to Brazil? BBC News, 12 November. http://news.bbc.co.uk/2/hi/business/8355343.stm (Accessed 2 November 2011).

Dussell, E. (1976) *Filosofía de la Liberación.* Mexico City, Mexico: Editorial Edicol.

Economist, The (2007) Out of the closet: And into politics. 8 March. www.economist.com/node/8819803 (Accessed 12 December 2011).

Edwards, M. (1989) The irrelevance of development studies. *Third World Quarterly* 11(1): 116–135.

Edwards, S. (2010) *Left Behind: Latin America and the False Promise of Populism.* Chicago: University of Chicago Press.

EFE (2008) En España hay 1.8 millones de inmigrantes latinoamericanos según informe. www.publico.es/agencias/efe/129158/en-espana-hay-1-8-millones-de-inmigrantes-latinoamericanos-segun-informe (Accessed 28 August 2011).

Elson, D. (1991) Structural adjustment: Its effect on women. In T. Wallace and C. March (eds) *Changing Perceptions: Writings on Gender and Development.* Oxford: Oxfam, pp. 39–53.

Elson, D. and Pearson, R. (1981) Nimble fingers make cheap workers: An analysis of women's employment in third world export manufacturing. *Feminist Review* 7: 87–107.

Encarnación, O. (2011) Latin America's gay rights revolution. *Journal of Democracy* 22(2): 104–119.

Ennis, S. R., Ríos-Vargas, M. and Albert, N. G. (2011) *The Hispanic Population: 2010.* Washington, DC: United States Census Bureau.

Escobar, A. (1995) *Encountering Development: The Making and Unmaking of the Third World.* Princeton, NJ: Princeton University Press.

Escobar, A. (2007) Worlds and knowledges otherwise. *Cultural Studies* 21(2–3): 179–210.

Escobar, A. (2008) *Territories of Difference: Place, Movements, Life, Redes.* Durham, NC: Duke University Press.

Escobar, A. and Alvarez, S. (eds) (1992) *The Making of Social Movements in Latin America: Identity, Strategy, and Democracy.* Boulder, CO: Westview Press.

Esteva, G. (1987) Regenerating people's space. *Alternatives* 12(1): 125–152.

Esteva, G. (2010) The Oaxaca commune and Mexico's coming insurrection. *Antipode* 42(4): 978–993.

Euraque, D. (1996) *Reinterpreting the Banana Republic: Region and State in Honduras 1870–1972.* Chapel Hill, NC: University of North Carolina Press.

Favell, A. (2010) State of mobile in Latin America: The latest stats, research and trends. *MobiThinking* 6 September. http://mobithinking.com/blog/latin_america_mobile_stats (Accessed 18 January 2012).

Fearnside, P. M. (2005) Deforestation in Brazilian Amazonia: History, rates and consequences. *Conservation Biology* 19(3): 680–688.

Feijoó, M. and Gogna, M. (1990) Women in the transition to democracy. In E. Jelin (ed.) *Women and Social Change in Latin America*. London: Zed Books, pp. 79–114.

Fernandes, S. (2010) *Who Can Stop the Drums? Urban Social Movements in Chávez's Venezuela*. Durham, NC: Duke University Press.

Fernández Jilberto, A. E. and Hogenboom, B. (2010a) Latin America and China: South–South relations in a new era. In A. E. Fernández Jilberto and B. Hogenboom (eds) *Latin America Facing China: South–South Relations beyond the Washington Consensus*. New York: Berghahn Books, pp. 1–32.

Fernández Jilberto, A. E. and Hogenboom, B. (2010b) Latin America – From Washington Consensus to Beijing Consensus? In A. E. Fernández Jilberto and B. Hogenboom (eds) *Latin America Facing China: South South Relations beyond the Washington Consensus*. New York: Berghahn Books, pp. 181–193.

Ffrench-Davis, R. (2010) *Economic Reforms in Chile: From Dictatorship to Democracy*. Basingstoke, UK: Palgrave Macmillan.

Field, L. W. (1999) *The Grimace of Macho Ratón: Artisans, Identity, and Nation in Late-Twentieth Century Western Nicaragua*. Durham, NC: Duke University Press.

Fischer, E. (2004) Beyond victimization: Maya movements in post-war Guatemala. In N. G. Postrero and L. Zamosc (eds) *The Struggle for Indigenous Rights in Latin America*. Eastbourne, UK: Sussex Academic Press, pp. 81–104.

Fisher, J. (1989) *Mothers of the Disappeared*. London: Zed Books.

Fisher, J. (1993) *Out of the Shadows: Women, Resistance and Politics in South America*. London: Latin America Bureau.

Fitz, D. (2007) Hybrids, biofuels and other false idols: What's being left out of solutions to fossil fuel? *Counterpunch* 14 February. www.counterpunch.org/2007/02/14/what-s-being-left-out-of-solutions-to-fossil-fuel (Accessed 12 November 2011).

Flusty, S. (2004) *De-Coca-Colonization: Making the Globe from the Inside Out*. London: Routledge.

Foley, E. (2012) Latino voters in election 2012 help sweep Obama to reelection. *Huffington Post* 7 November. www.huffingtonpost.com/2012/11/07/latino-voters-election-2012_n_2085922.html (Accessed 5 December 2012).

Foley, N. (2010) *Quest for Equality: The Failed Promise of Black–Brown Solidarity*. Cambridge, MA: Harvard University Press.

Fowler, W. (2007) *Santa Anna of Mexico*. Lincoln, NE: University of Nebraska Press.

Fox, E. and Waisbord, S. (2002a) Introduction. In E. Fox and S. Waisbord (eds) *Latin Politics, Global Media*. Austin, TX: University of Texas Press, pp. ix–xxii.

Fox, E. and Waisbord, S. (eds) (2002b) *Latin Politics, Global Media*. Austin, TX: University of Texas Press.

Franco, J. (2004) Plotting women: Popular narratives for women in the US and in Latin America. In A. del Sarto, A. Ríos and A. Trigo (eds) *The Latin American Cultural Studies Reader*. Durham, NC: Duke University Press, pp. 183–202.

Franko, P. (2007) *The Puzzle of Latin American Economic Development*, 3rd edn. Lanham, MD: Rowman & Littlefield.

French, J. H. (2009) *Legalizing Identities: Becoming Black or Indian in Brazil's Northeast*. Chapel Hill, NC: University of North Carolina Press.

Froehling, O. (1997) The cyberspace "war of ink and Internet" in Chiapas, Mexico. *Geographical Review* 87(2): 291–307.

FSLN (1969) *Programa Sandinista*. Reproduced in D. Gilbert and D. Block (1990) *Sandinistas: Key Documents/Documentos Claves*. Ithaca, NY: Latin American Studies Program, Cornell University, pp. 3–21.

Galeano, E. (1971) *Las Venas Abiertas de América Latina.* México Siglo XXI, Mexico City, Mexico.

García Canclini, N. (2001) *Consumers and Citizens: Globalization and Multicultural Conflicts.* Minneapolis, MN: University of Minnesota Press.

García Canclini, N. (2002) *Latinoamericanos Buscando Lugar en este Siglo.* Buenos Aires, Argentina: Paidós.

Garibay, C., Boni, A., Panico, F., Urquijo, P. and Klooster, D. (2011) Unequal partners, unequal exchange: Goldcorp, the Mexican state, and campesino dispossession at the Peñasquito goldmine. *Journal of Latin American Geography* 10(2): 153–176.

Garwood, S. (2002) Working to death: Gender, labour, and violence in Ciudad Juárez, Mexico. *Peace, Conflict and Development* 2: 1–23.

Gates, H. L., Jr (2011) *Black in Latin America.* New York: New York University Press.

Ghosh, J. (2012) Could Ecuador be the most radical and exciting place on Earth? *The Guardian* 19 January. www.guardian.co.uk/commentisfree/cifamerica/2012/jan/19/ecuador-radical-exciting-place (Accessed 20 February 2012).

Gill, L. (2004) *The School of the Americas: Military Training and Political Violence.* Durham, NC: Duke University Press.

Gill, R. B. (2000) *The Great Maya Droughts: Water, Life, and Death.* Albuquerque, NM: University of New Mexico Press.

Ginsburg, F. (2002). Screen memories: Resignifying the traditional in indigenous media. In F. D. Ginsburg, L. Abu-Lughod and B. Larkin (eds) *Media Worlds: Anthropology on New Terrain.* Berkeley, CA: University of California Press, pp. 39–57.

Glynn, K. and Cupples, J. (2011) Indigenous mediaspace and the production of (trans)locality on Nicaragua's Mosquito Coast. *Television and New Media* 12(2): 101–135.

Godoy, E. (2009) Mexico: Black minority invisible in bicentennial plans. IPS, 2 October. www.ipsnews.net/2009/10/mexico-black-minority-invisible-in-bicentennial-plans (Accessed 14 August 2012).

González, A. (2011) Landscape and settlement patterns. In J. K. Black (ed.) *Latin America: Its Problems and Its Promise: A Multidisciplinary Introduction.* Boulder, CO: Westview Press, pp. 23–38.

González de la Rocha, M. (ed.) (1999) *Divergencias del modelo tradicional: Hogares de jefatura femenina en América Latina.* Mexico City, Mexico: CIESAS.

Goodman, D. and Redclift, M. (1991) Introduction. In D. Goodman and M. Redclift (eds) *Environment and Development in Latin America: The Politics of Sustainability.* Manchester, UK: Manchester University Press, pp. 1–23.

Goodman, J. (2007) Greenwashing energy crops: Biofuels, the biggest scam going. *Counterpunch* 28 December. www.counterpunch.org/goodman12282007.html (Accessed 12 November 2011).

Graham, R. (1990) Introduction. In R. Graham (ed.) *The Idea of Race in Latin America 1870–1940.* Austin, TX: University of Texas Press, pp.1–5.

Green, D. (2006) *Faces of Latin America,* 3rd edn. London: Latin America Bureau.

Greenpeace (2009) Cattle ranching expansion in the Brazilian Amazon. www.greenpeace.org/international/Global/international/planet-2/report/2009/1/cattle-ranching-expansion-in-t.pdf (Accessed 12 November 2009).

Greenslade, R. (2006) 100 journalists murdered in Colombia since 1980. *The Guardian* 7 July. www.guardian.co.uk/media/greenslade/2006/jul/07/100journalistsmurderedinco (Accessed 12 January 2012).

Grillo, C. (1995) Brasil quer ser chamado moreno y só 39% se autodefinem como brancos. *Folha de São Paulo* 25 June. http://almanaque.folha.uol.com.br/racismo05.pdf (Accessed 26 July).

Guevara, A. (2008) Pesticide, performance, protest: Theatricality of flesh in Nicaragua. *Intensions* 1: 1–23.

Guthmann, J. (1995) Epidemic cholera in Latin America: Spread and routes of transmission. *Journal of Tropical Medicine and Hygiene* 98(6): 419–427.

Gutierrez, G. (1974) *A Theology of Liberation: History, Politics, and Salvation.* London: SCM Press.

Gutmann, M. C. (1996) *The Meanings of Macho: Being a Man in Mexico City.* Berkeley, CA: University of California Press.

Guzman Bouvard, M. (1994) *Revolutionizing Motherhood: The Mothers of the Plaza de Mayo.* Wilmington, DE: Scholarly Resources.

Hale, C. (1992) *Resistance and Contradiction: Miskitu Indians and the Nicaraguan State 1894–1987.* Stanford, CA: Stanford University Press.

Hale, C. (2004) Rethinking indigenous politics in the era of the "indio permitido". *NACLA Report on the Americas* 38(2): 16–21.

Hale, C. (2006) Activist research v. cultural critique: Indigenous land rights and the contradictions of politically engaged anthropology. *Cultural Anthropology* 21(1): 96–120.

Hall, S. (1996) Gramsci's relevance for the study of race and ethnicity. In D. Morley and K. Chen (eds) *Stuart Hall: Critical Dialogues in Cultural Studies.* London: Routledge, pp. 411–440.

Halleck, D. (1994) Zapatistas on-line. *NACLA Report on the Americas* 38(2): 30–32.

Hamilton, A. (2009) Yoani Sanchez, Cuba's popular blogger, has been beaten up for describing life. *The Telegraph* 28 November. www.telegraph.co.uk/news/worldnews/centralamericaandthecaribbean/cuba/6678937/Yoani-Sanchez-Cubas-popular-blogger-has-been-beaten-up-for-describing-life.html (Accessed 20 January 2012).

Hamilton, R. G. (2008) European transplants, Amerindian in-laws, African settlers, Brazilian creoles: A unique colonial and postcolonial condition in Latin America. In W. D. Mignolo, I. Silverblatt and S. Saldívar-Hull (eds) *Coloniality at Large: Latin America and the Postcolonial Debate.* Durham, NC: Duke University Press, pp. 113–129.

Hammond, J. L. (2011) The resource curse and oil revenues in Angola and Venezuela. *Science and Society* 75(3): 348–378.

Harris, N. (2007) *Environmental Issues in the Cities of the Developing World.* Working Paper 20, Development Planning Unit, University College London. www.ucl.ac.uk/dpu/k_s/publications/working_papers/f-j/WP20.pdf (Accessed 12 November 2011).

Harris, O. (1981) Households as natural units. In K. Young, C. Wolkowitz and R. McCullagh (eds) *Of Marriage and the Market.* London: CSE, pp. 48–67.

Hawley, S. (1997) Protestantism and indigenous mobilisation: The Moravian Church among the Miskitu Indians of Nicaragua. *Journal of Latin American Studies* 29(1): 111–129.

Helms, M. (1971) *Asang: Adaptations to Culture Contact in a Miskito Community.* Gainesville, FL: University of Florida Press.

Hewitt, K. (1995) Excluded perspectives in the social construction of disaster. *International Journal of Mass Emergencies and Disasters* 13(3): 317–339.

Holt, T. C. (2003) Foreword: The first new nations. In N. P. Appelbaum, A. S. Macpherson and K. A. Rosemblatt (eds) *Race and Nation in Modern Latin America.* Chapel Hill, NC: University of North Carolina Press, pp. vii–xvi.

Howard, A. and Dangl, B. (2007) The multinational beanfield war: Soy cultivation spells doom for Paraguayan campesinos. *In These Times* 12 April. www.inthesetimes.com/article/3093 (Accessed 6 November 2011).

Howe, C. (2008) Spectacles of sexuality: Televisionary activism in Nicaragua. *Cultural Anthropology* 23(1): 48–84.

Hoy, P. (1998) *Players and Issues in International Aid.* West Hartford, CT: Kumarian Press.

Hughes, S. and Gil, J. (2004) The civic transformation of Mexican newspapers. *NACLA Report on the Americas* 37(4): 26–28.

Humphreys, D. (1996) *Forest Politics: The Evolution of International Cooperation.* London: Earthscan.

Irazábal, C. (2005) *City Making and Urban Governance in the Americas: Curitiba and Portland.* Farnham, UK: Ashgate.

Irwin, R. M. (2009) Memín Pinguín: Líos gordos con los gringos. In H. Fernández L'Hoeste and J. Poblete (eds) *Redrawing the Nation: National Identity in Latin(o) American Comics.* New York: Palgrave Macmillan, pp. 111–130.

Jaquette, J. (2009) Introduction. In J. Jaquette (ed.) *Feminist Agendas and Democracy in Latin America.* Durham, NC: Duke University Press, pp. 1–18.

Johnson, K. (1985) Nicaraguan culture: Unleashing creativity. *NACLA Report on the Americas* 19(5): 8–11.

Kaimowitz, D. (2008) The prospects for reduced emissions from deforestation and degradation (REDD) in Mesoamerica. *International Forestry Review* 10(3): 485–495.

Kaimowitz, D., Mertens, B., Wunder, S. and Pacheco, P. (2004) *Hamburger Connection Fuels Amazon Destruction: Cattle Ranching in Brazil's Amazon.* Technical report. Bogor, Indonesia: Center for International Forest Research. www.cifor.cgiar.org/publications/pdf_files/media/Amazon.pdf (Accessed 11 July 2011).

Kampwirth, K. (2002) *Women and Guerrilla Movements: Nicaragua, El Salvador, Chiapas, Cuba.* University Park, PA: Pennsylvania State University Press.

Kampwirth, K. (2004) *Feminism and the Legacy of Revolution: Nicaragua, El Salvador, Chiapas.* Athens, OH: Ohio University Press.

Katzew, A. (2011) Shut up! Representations of the Latino/a body in Ugly Betty and their educational implications. *Latino Studies* 9(2–3): 300–320.

Kaufman, T. (1994a) The native languages of Meso-America. In C. Moseley and R. E. Asher (eds) *Atlas of the World's Languages.* London: Routledge, pp. 34–41.

Kaufman, T. (1994b) The native languages of South America. In C. Moseley and R. E. Asher (eds) *Atlas of the World's Languages.* London: Routledge, pp. 46–76.

Keck, M. (1995) Parks, people and power: The shifting terrain of environmentalism. *NACLA Report on the Americas* 28(5): 36–41.

Klein, N. (2007) *The Shock Doctrine: The Rise of Disaster Capitalism.* New York: Metropolitan Books.

Knight, A. (1990) Racism, revolution, and indigenismo: Mexico, 1910–1940. In R. Graham (ed.) *The Idea of Race in Latin America, 1870–1940.* Austin, TX: University of Texas Press.

Kowal, D. M. (2002) Digitizing and globalizing indigenous voices: The Zapatista movement. In G. Elmer (ed.) *Critical Perspectives on the Internet.* Lanham, MD: Rowman & Littlefield, pp. 105–121.

Krantz, R. (2003) Cycles of reform in Porto Alegre and Madison. In A. Fund and E. O. Wright (eds) *Deepening Democracy: Institutional Innovations on Empowered Participatory Governance.* London: Verso, pp. 225–236.

Lamrani, S. (2009) Las contradicciones de la bloguera cubana Yoani Sánchez. *Rebelión* 24 November. http://www.rebelion.org/noticias/2009/11/95809.pdf (Accessed 18 January 2012).

Lamrani, S. (2010) Yoani Sanchez interviewed by Salim Lamrani. *Cuba Internal Reform* 27 April. http://internalreform.blogspot.com/2010/04/yoani-sanchez-interviewed-by-salim.html (Accessed 18 January 2012).

Lancaster, R. N. (1992) *Life is Hard: Machismo, Danger and the Intimacy of Power in Nicaragua.* Berkeley, CA: University of California Press.

Lane, J. (2003) Digital Zapatistas. *The Drama Review* 47(2): 129–144.

Laurell, A. C. (2000) Structural adjustment and the globalization of social policy in Latin America. *International Sociology* 15(2): 306–325.

Lauría, C. (2004) Reporters under fire in Colombia. *NACLA Report on the Americas* 37(4): 36–38.

La Voz de Los de Abajo (2012) Delegation report: Standard Fruit uses the army & police to attack campesinos. *Honduras Resists/Honduras Resiste* 12 February. http://hondurasresists.blogspot.co.nz/2012/02/delegation-report-standard-fruit-uses.html (Accessed 20 February 2012).

Laurie, N., Dwyer, C., Holloway, S. and Smith, F. (1999) *Geographies of New Femininities.* Harlow, UK: Longman.

Laurie, N., Andolina, R. and Radcliffe, S. (2002) The excluded "indigenous"? The implications of multi-ethnic politics for water reform in Bolivia. In R. Sieder (ed.) *Multiculturalism in Latin America: Indigenous Rights, Diversity and Democracy.* Basingstoke, UK: Palgrave Macmillan, pp. 252–276.

Lecount, C. (1999) Carnival in Bolivia: Devils dancing for the virgin. *Western Folklore* 58(3/4): 231–252.

LeGrand, C. (1998) Living in Macondo: Economy and culture in a United Fruit Company banana enclave in Colombia. In G. M. Joseph, C. LeGrand and R. D. Salvatore (eds) *Close Encounters of Empire: Writing the Cultural History of US–Latin American Relations.* Durham, NC: Duke University Press, pp. 333–368.

Leóns, M. B. and Sanabria, H. (1997) Coca and cocaine in Bolivia: Reality and policy illusion. In M. B. Leóns and H. Sanabria (eds) *Coca, Cocaine and the Bolivian Reality.* New York: SUNY Press, pp. 1–46.

Libendinsky, N. (1999) Fútbol, polo y tango, los tres pilares de la identidad argentina. *La Nación* 17 January. www.lanacion.com.ar/nota.asp?nota_id=124858 (Accessed 15 January 2012).

Lindsay, C. (2007) *The Latin American Community 2001.* Ottawa, Canada: Statistics Canada. www.statcan.gc.ca/pub/89-621-x/89-621-x2007008-eng.pdf (Accessed 28 August 2011).

Linneker, B. and McIlwaine, C. (2011) *Estimating the Latin American Population of London from Official Data Sources.* London: School of Geography, Queen Mary University of London. ww.geog.qmul.ac.uk/docs/research/latinamerican/48640.pdf (Accessed 28 August 2011).

Liverman, D. A. and Vilas, S. (2006) Neoliberalism and the environment in Latin America. *Annual Review of Environmental Resources* 31: 327–363.

Livingstone, G. (2009) *America's Backyard: The United States and Latin America from the Monroe Doctrine to the War on Terror.* London: Zed Books.

López, A. M. (1995) Our welcomed guests: Telenovelas in Latin America. In

R. C. Allen (ed.) *To Be Continued: Soap Operas Around the World.* London: Routledge, pp. 256–275.

Lopez-Gamundi, P. and Hanks, W. (2011) A land-grabber's loophole. *COHA* 8 August. www.coha.org/a-land-grabbers-loophole (Accessed 24 August 2011).

López Vigil, J. I. (1995) *Rebel Radio: The Story of El Salvador's Radio Venceremos.* Willimantic, CT: Curbstone Press.

Lovell, P. A. (1999) Development and the persistence of racial inequality in Brazil: 1950–1991. *Journal of Developing Areas* 33: 395–418.

McCarthy, J. and Prudham, S. (2004) Neoliberal nature and the nature of neoliberalism. *Geoforum* 35: 275–283.

McClenaghan, S. (1997) Women, work and empowerment: Romanticizing the reality. In E. Dore (ed.) *Gender Politics in Latin America: Debates in Theory and Practice.* New York: Monthly Review Press, pp. 19–35.

McEwan, C. (2009) *Postcolonialism and Development.* London: Routledge.

McGuinness, A. (2003) Searching for "Latin America": Race and sovereignty in the Americas in the 1850s. In N. P. Appelbaum, A. S. Macpherson and K. A. Rosemblatt (eds) *Race and Nation in Modern Latin America.* Chapel Hill, NC: University of North Carolina Press, pp. 87–107.

Maldonado R., Bajuk, N. and Hayem, N. (2012) *Remittances to Latin America and the Caribbean: Regaining Growth.* Washington, DC: Multilateral Investment Fund, Inter-American Development Bank.

Manzano, O. and Monaldi, F. (2008) The political economy of oil production in Latin America. *Economía* 9(1): 59–103.

Martí, J. (1977[1891]) Our America. In P. S. Foner (ed.) *Our America by José Martí: Writings on Latin America and the Struggle for Cuban Independence.* New York: Monthly Review Press, pp. 84–94.

Martí i Puig, S. (2010) The emergence of indigenous movements in Latin America and their impact on the Latin American political scene: Interpretive tools at the local and global levels. *Latin American Perspectives* 37(6): 74–92.

Martín Barbero, J. (2004a) *De los medios a las mediaciones: Comunicación, cultura y hegemonía.* Barcelona, Spain: Editorial Gustavo Gili.

Martín Barbero, J. (2004b) A nocturnal map to explore a new field. In A. del Sarto, A. Ríos and A. Trigo (eds) *The Latin American Cultural Studies Reader.* Durham, NC: Duke University Press, pp. 310–328.

Martínez, I. (2005) Romancing the globe. *Foreign Policy* 9 November. www.foreignpolicy.com/articles/2005/11/09/romancing_the_globe?page=0,0 (Accessed 15 January 2012).

Menchú, R. and Burgos-Debray, E. (1984) *I, Rigoberta Menchú: An Indian Woman in Guatemala.* London: Verso.

Merrell, F. (2004) *Complementing Latin America's Borders.* West Lafayette, IN: Purdue University Press.

Mignolo, W. D. (2005) *The Idea of Latin America.* Malden, MA: Blackwell.

Mignolo, W. D. (2007a) Coloniality of power and decolonial thinking. *Cultural Studies* 21(2–3): 155–167.

Mignolo, W. D. (2007b) Delinking: The rhetoric of modernity, the logic of coloniality and the grammar of de-coloniality. *Cultural Studies* 21(2–3): 449–514.

Milton, K. (1993) *Environmentalism: The View from Anthropology.* London: Routledge.

Moberg, M. and Striffler, S. (2003) Introduction. In S. Striffler and M. Moberg (eds)

Banana Wars: Power, Production and History in the Americas. Durham, NC: Duke University Press, pp. 1–19.

Molyneux, M. (1986) Mobilization without emancipation? Women's interests, state, and revolution. In R. F. Fagan, C. D. Deere and J. L. Coraggio (eds) *Transition and Development: Problems of Third World Socialism.* New York: Monthly Review Press, pp. 280–302.

Monahan, J. (2005) Soybean fever transforms Paraguay. BBC News, 6 June. http://news.bbc.co.uk/2/hi/business/4603729.stm (Accessed 6 November 2011).

Montenegro, S. (1997) *La Revolución Simbólica Pendiente: Mujeres, Medios de Comunicación y Política.* Managua, Nicaragua: CINCO.

Montoya Tellería, O. (1998) *Nadando contra Corriente: Buscando Pistas para Prevenir la Violencia Masculina en las Relaciones de Pareja.* Managua, Nicaragua: Puntos de Encuentro.

Moore, D. S. (1996) Marxism, culture and political ecology: Environmental struggles in Zimbabwe's eastern highlands. In R. Peet and M. Watts (eds) *Liberation Ecologies: Environment, Development, Social Movements.* London: Routledge, pp. 125–147.

Moseley, T. (2010) Four capital cities sign up to 10:10. *The Guardian* 14 October. www.guardian.co.uk/environment/2010/oct/14/10 10-capital-cities? (Accessed 12 November 2011).

Moser, C. O. N. (1993) Adjustment from below: Low-income women, time and the triple role in Guayaquil, Ecuador. In S. A. Radcliffe and S. Westwood (eds) *Viva: Women and Popular Protest in Latin America.* London: Routledge, pp. 173–196.

Mowforth, M. (1999) Hurricane Mitch – a natural disaster? *Central America Report.* Spring.

Muntaner, C., Guerra Salazar, R. M., Rueda, S. and Armada, F. (2006) Challenging the neoliberal trend: The Venezuelan health care reform alternative. *Canadian Journal of Public Health* 97(6): 19–24.

Murguialday, C. (1990) *Nicaragua, revolución y feminismo (1977–89).* Madrid, Spain: Ed. Revolución.

Mychalejko, C. (2008) Ecuador's constitution gives rights to nature. *Upside Down World* 25 September. http://upsidedownworld.org/main/content/view/1494/49 (Accessed 9 November 2011).

NACLA (1996) Gaining ground: The indigenous movement in Latin America. *NACLA Report on the Americas* 29(5): 14.

Neate, P. and Platt, D. (2010) *Culture is our Weapon: Making Music and Changing Lives in Rio de Janeiro.* London: Penguin.

Neilson, B. (2004) *Free Trade in the Bermuda Triangle – and Other Tales of Counter-Globalization.* Minneapolis, MN: University of Minnesota Press.

Niehaus, T. (1994) Interview with Indiana Acevedo, July 12 1992, Managua, Nicaragua. In G. M. Yeager (ed.) *Confronting Change, Challenging Tradition: Women in Latin American History.* Wilmington, DE: Scholarly Resources, pp. 211–218.

Nietschmann, B. (1994) Defending the Miskito Reefs with maps and GPS: Mapping with sail, scuba and satellite. *Cultural Survival* 18(4). www.culturalsurvival.org/ourpublications/csq/article/defending-miskito-reefs-with-maps-and-gps-mapping-with-sail-scuba-and-sa (Accessed 12 January 2012).

Niezen, R. (2003) *The Origins of Indigenism: Human Rights and the Politics of Identity.* Berkeley, CA: University of California Press.

Noonan, R. K. (1995) Women against the state: Political opportunities and collective action frames in Chile's transition to democracy. *Sociological Forum* 10(1): 81–111.

Nordhaus, T. and Shellenberger, M. (2007) *Break Through: From the Death of Environmentalism to the Politics of Possibility.* New York: Houghton Mifflin.

O'Brien, P. J. (1991) Debt and sustainable development. In D. Goodman and M. Redclift (eds) *Environment and Development in Latin America: The Politics of Sustainability.* Manchester, UK: Manchester University Press, pp. 24–47.

O'Neill, J. (2001) Building better global economic BRICs. *Goldman Sachs Global Economics Paper*, 66. www.goldmansachs.com/our-thinking/brics/brics-reports-pdfs/build-better-brics.pdf (Accessed 25 March 2012).

Ortiz, F. (1995[1940]) *Cuban Counterpoint: Tobacco and Sugar.* Durham, NC: Duke University Press.

O'Shaughnessy, H. (2007) Poisoned city fights to save its children. *The Observer* 12 August.

Oxfam Policy Department (1995) *A Case for Reform: Fifty Years of the IMF and World Bank.* Oxford: Oxfam.

Pachico, E. (2009) Rural revolution in Colombia goes digital. *Upside Down World* 9 September. http://upsidedownworld.org/main/colombia-archives-61/2101-rural-revolution-in-colombia-goes-digital (Accessed 16 January 2012).

Palaversich, D. (1999) Caught in the act: Social stigma, homosexual panic and violence in Latin American writing. *Chasqui: Revista de Literatura Latinoamericana* 28(2): 60–75.

Palhares, G. (2011) Lagarde meets with Rousseff and Mantega as part of her visit to Latin America to negotiate Brazil's lending to the IMF. *COHA* 8 December. www.coha.org/lagarde-meets-with-rousseff-and-mantega-as-part-of-her-visit-to-latin-america-to-negotiate-brazil%E2%80%99s-lending-to-the-imf (Accessed 14 December 2011).

Panizza, F. (2009) *Contemporary Latin America: Development and Democracy beyond the Washington Consensus.* London: Zed Books.

Paz, O. (1950) *El Laberinto de la Soledad.* Mexico City, Mexico: Cuadernos Americanos.

Pearce, J. (2007) Oil and armed conflict in Casanare, Colombia: Complex contexts and contingent moments. In M. Kaldor, T. L. Karl and Y. Said (eds) *Oil Wars.* London: Pluto Press, pp. 225–273.

Peet, R. (2003) *Unholy Trinity: The IMF, the World Bank and the WTO.* London: Zed Books.

Peet, R. (2007) *Geography of Power: The Making of Global Economic Policy.* London: Zed Books.

Pellow, D. N. (2007) *Resisting Global Toxics: Transnational Movements for Environmental Justice.* Cambridge, MA: MIT Press.

Perreault, T. (2001) Developing identities: Indigenous mobilization, rural livelihoods, and resource access in Ecuadorian Amazonia. *Ecumene* 8(4): 381–413.

Perreault, T. (2005) State restructuring and the scale politics of rural water governance in Bolivia. *Environment and Planning A* 37(2): 263–284.

Peters, P. (1995) The use and abuse of the concept of "female-headed households" in research on agrarian transformation and policy. In D. F. Bryceson (ed.) *Women Wielding the Hoe: Lessons from Rural Africa for Feminist Theory and Development.* Oxford: Berg, pp. 93–108.

Petras, J. and Veltmeyer, H. (2011) *Social Movements in Latin America: Neoliberalism and Popular Resistance.* New York: Palgrave Macmillan.

Phillips, T. (2011) Brazil census shows African-Brazilians in the majority for the first time. *The Guardian* 17 November. www.guardian.co.uk/world/2011/nov/17/brazil-census-african-brazilians-majority? (Accessed 9 December 2011).

Pires de Rio Caldeira, T. (1990) Women, daily life and politics. In E. Jelin (ed.) *Women and Social Change in Latin America.* London: Zed Books, pp. 47–78.

Pitt-Rivers, J. (1967) Race, color, and class in Central America and the Andes. *Daedalus* 96(2): 542–559.

Postero, N. G. (2004) Articulation and fragmentation: Indigenous politics in Bolivia. In N. G. Postrero and L. Zamosc (eds) *The Struggle for Indigenous Rights in Latin America.* Eastbourne, UK: Sussex Academic Press, pp. 189–216.

Postero, N. G. and Zamosc, L. (2004) Indigenous movements and the Indian question in Latin America. In N. G. Postrero and L. Zamosc (eds) *The Struggle for Indigenous Rights in Latin America.* Eastbourne, UK: Sussex Academic Press, pp. 1–31.

Potter, G. A. (2000) *Deeper than Debt: Economic Globalisation and the Poor.* London: Latin America Bureau.

Prandi, R. (2008) Religions and cultures: Religious dynamics in Latin America. *Social Compass* 55(3): 264–274.

Prebisch, R. (1950) *The Economic Development of Latin America and Its Principal Problems.* New York: United Nations.

Puar, J. K. (1996) Nicaraguan women, resistance, and the politics of aid. In H. Afshar (ed.) *Women and Politics in the Third World.* London: Routledge, pp. 73–92.

Quijano, A. (2005) The challenge of the "indigenous movement" in Latin America. *Socialism and Democracy* 19(3): 55–78.

Quijano, A. (2007) Coloniality and modernity/rationality. *Cultural Studies* 21(2–3): 168–178.

Rabinovitch, J. and Leitman, J. (1996) Urban planning in Curitiba. *Scientific American* 274(3): 46–53.

Radcliffe, S. A. and Westwood, S. (1996) *Remaking the Nation: Place, Identity and Politics in Latin America.* London: Routledge.

Rahnema, M. (1997) Towards post-development: Searching for signposts, a new language and new paradigms. In M. Rahnema and V. Bawtree (eds) *The Post-Development Reader.* London: Zed Books, pp. 377–403.

Ramos, A. R. (1998) *Indigenism: Ethnic Politics in Brazil.* Madison, WI: University of Wisconsin Press.

Ramos, A. R. (2006) The commodification of the Indian. In D. A. Posey (ed.) *Human Impacts on Amazonia: The Role of Traditional Ecological Knowledge in Conservation and Development in Brazil.* New York: Columbia University Press.

Ramos, A. R., Guerreiro Osório, R. and Pimenta, J. (2009) Indigenising Development. In A. R. Ramos, R. Guerreiro Osório and J. Pimenta (eds) Indigenising development, *Poverty in Focus* 17. Brasilia, Brazil: International Policy Centre for Inclusive Growth, UNDP. www.ipc-undp.org/pub/IPCPoverty InFocus17.pdf (Accessed 14 December 2011).

Randall, M. (1981) *Sandino's Daughters: Testimonies of Nicaraguan Women in Struggle.* Vancouver, Canada: New Star.

Randall, M. (1994) *Sandino's Daughters Revisited: Feminism in Nicaragua.* New Brunswick, NJ: Rutgers University Press.

Rathgeber, E. (1990) WID, WAD, GAD: Trends in research and practice. *Journal of Developing Areas* 24: 489–502.

Raynolds, L. (1998) Harnessing women's work: Restructuring agricultural and industrial labor forces in the Dominican Republic. *Economic Geography* 74(2): 149–169.

redd-monitor.org (2011) REDD: An introduction. www.redd-monitor.org/redd-an-introduction (Accessed 7 September 2012).

Remedi, G. A. (2004) The production of local public spheres: Community radio stations. In A. del Sarto, A. Ríos and A. Trigo (eds) *The Latin American Cultural Studies Reader.* Durham, NC: Duke University Press, pp. 513–534.

Rendell, M. (2011) *Salsa for People Who Probably Shouldn't.* Edinburgh, UK: Mainstream.

Restall, M. (2004) *Seven Myths of the Spanish Conquest.* Oxford: Oxford University Press.

Revkin, A. (2004) *The Burning Season: The Murder of Chico Mendes and the Fight for the Amazon Rain Forest.* Washington, DC: Island Press.

Richards, K. (2005) Theater under dictatorship. In L. Shaw and S. Dennison (eds) *Pop Culture Latin America: Media, Arts, and Lifestyle.* Santa Barbara, CA: ABC Clio, pp. 121–125.

Rivero, Y. M. (2003) The performance and reception of televisual "ugliness" in *Yo soy Betty la fea. Feminist Media Studies* 3(1): 65–81.

Roberts, J. T. and Thanos, D. T. (2003) *Trouble in Paradise: Globalization and Environmental Crises in Latin America.* London: Routledge.

Rockwell, R. (2002) Corruption and calamity: Limiting ethical journalism in Mexico and Central America. *Global Media Journal* 1(1): 1–13.

Rockwell, R. and Janus, N. (2002) The triumph of the media elite in postwar Central America. In E. Fox and S. Waisbord (eds) *Latin Politics, Global Media.* Austin, TX: University of Texas Press, pp. 47–68.

Roddick, A. (2001) Sesame Feat. *Geographical* 73(6): 30–35.

Rosin, E. (2005) *Drugs and Democracy in Latin America: The Impact of US Policy.* Boulder, CO: Lynne Rienner.

Rostow, W. (1960) *The Stages of Economic Growth: A Non-Communist Manifesto.* Cambridge: Cambridge University Press.

Routledge, P. (1998) Going globile: Spatiality, embodiment, and mediation in the Zapatista insurgency. In G. Ó Tuathail and S. Dalby (eds) *Rethinking Geopolitics.* London: Routledge, pp. 240–260.

Rowe, W. and Shelling, V. (1991) *Memory and Modernity: Popular Culture in Latin America.* London: Verso.

Rushe, D. and Carroll, R. (2011) Chevron fined $8bn over Amazon "contamination". *The Guardian* 14 February. www.guardian.co.uk/business/2011/feb/14/chevron-contaminate-ecuador? (Accessed 11 November).

Sackur, S. (2012) Honduras counts the human rights cost of America's war on drugs. *The Guardian* 15 July. www.guardian.co.uk/world/2012/jul/15/honduras-human-rights-war-drugs? (Accessed 23 August 2012).

Safa, H. I. and Flora, C. B. (1992) Production, reproduction, and the polity: Women's strategic and practical gender issues. In A. Stepan (ed.) *Americas: New Interpretive Essays.* Oxford: Oxford University Press, pp. 109–136.

Salazar, J. F. (2009) Self-determination in practice: The critical making of indigenous media. *Development in Practice* 19(4–5): 504–513.

Salazar, J. F. and Córdova, A. (2008) Imperfect media and the poetics of indigenous video in Latin America. In P. Wilson and M. Stewart (eds) *Global Indigenous*

Media: Cultures, Poetics, and Politics. Durham, NC: Duke University Press, pp. 39–57.

Salgado, S. (2003) *The Teatro Solís: 150 Years of Opera, Concert and Ballet in Montevideo.* Middletown, CT: Wesleyan University Press.

Sami, F. (2011) Flow of Chinese investments continues to aid Brazil's ascendency. *Council on Hemispheric Affairs* 12 October. www.coha.org/flow-of-chinese-investments-continue-to-aid-brazil%E2%80%99s-ascendency (Accessed 20 February 2012).

Sánchez, Y. (2011) *Havana Real: One Woman Fights to Tell the Truth about Cuba Today.* Brooklyn, NY: Melville House.

Sandels, R. (2011) An oil-rich Cuba. *Monthly Review* 63(4): 40–45.

Sara-Lafosse, V. (1989) Los comedores y la promoción de la mujer. In N. M. Galer and P. Nuñez Carvallo (eds) *Mujer y comedores populares.* Lima, Peru: SEPADE, pp. 187–211.

Sawyer, S. (1997) The 1992 Indian mobilization in lowland Ecuador. *Latin American Perspectives* 24(3): 65–82.

Sawyer, S. (2004) *Crude Chronicles: Indigenous Politics, Multinational Oil and Neoliberalism in Ecuador.* Durham, NC: Duke University Press.

Schirmer, J. G. (1988) "Those who die for life cannot be called dead": Women and human rights protest in Latin America. *Harvard Human Rights Yearbook* 1: 41–76.

Schirmer, J. G. (1993) The seeking of truth and the gendering of consciousness: The CoMadres of El Salvador and the CONAVIGUA widows of Guatemala. In S. A. Radcliffe and S. Westwood (eds) *Viva: Women and Popular Protest in Latin America.* London: Routledge, pp. 30–64.

Schiwy, F. (2003) Decolonizing the frame: Indigenous video in the Andes. *Framework* 44(1): 116–132.

Schwerin, K. H. (2011) The Indian populations of Latin America. In J. K. Black (ed.) *Latin America: Its Problems and Its Promise: A Multidisciplinary Introduction.* Boulder, CO: Westview Press, pp. 39–55.

Shankland, A. (1993) The natives are friendly! *New Internationalist* 245(July). www.newint.org/issue245/natives.htm (Accessed 14 June 2002).

Shiva, V. (2002) *Water Wars: Privatization, Pollution and Profit.* Cambridge, MA: South End Press.

Sinclair, J. (2004) The globalization of Latin American media. *NACLA Report on the Americas* 37(4): 15–19.

Singhal, A. and Rogers, E. M. (1999) *Education–Entertainment: A Communication Strategy for Social Change.* London: Routledge.

Skidmore, T. (1993) Bi-racial U.S.A. vs. multi-racial Brazil: Is the contrast still valid? *Journal of Latin American Studies* 25: 373–386.

Skidmore, T. E. and Smith, P. H. (2005) *Modern Latin America*, 6th edn. Oxford: Oxford University Press.

Slovo, G. (2012) 'In Mexico, reporters are hunted like rabbits'. *The Guardian* 3 February. www.guardian.co.uk/books/2012/feb/03/author-author-gillian-slovo (Accessed 4 February 2012).

Smith, M. P. (2001) *Transnational Urbanism: Locating Globalization.* Malden, MA: Blackwell.

Solnit, R. (2010) *A Paradise Built in Hell: The Extraordinary Communities that Arise in Disaster.* New York: Penguin Books.

Soltis, K. (2011a) A victory for gay rights in Latin America. *COHA* 25 May.

www.coha.org/latin-america-progresses-forward-a-victory-for-gay-rights (Accessed 12 December 2011).

Soltis, K. (2011b) WikiLeaks cables show Haiti as pawn in U.S. foreign policy. *COHA* 27 July. www.coha.org/wikileaks-cables-show-haiti-as-pawn-in-u-s-foreign-policy (Accessed 19 August 2011).

Soluri, J. (2005) *Banana Cultures: Agriculture, Consumption and Environmental Change in Honduras and the United States.* Austin, TX: University of Texas Press.

Soong, R. (1999) Telenovelas in Latin America. *Zona Latina* 11 August. www.zonalatina.com/Zldata70.htm (Accessed 12 January 2012).

Sparr, P. (ed.) (1994) *Mortgaging Women's Lives: Feminist Critiques of Structural Adjustment.* London: Zed Books.

Standing, G. (1989) Global feminization through flexible labour. *World Development* 17: 1077–1095.

Stavenhagen, R. (1994) Challenging the nation-state in Latin America. In J. I. Domínguez (ed.) *Race and Ethnicity in Latin America.* New York: Garland, pp. 329–348.

Stea, D. and Lewis, G. S. (2011) Harmonizing and disharmonizing human and natural environments. In J. K. Black (ed.) *Latin America: Its Problems and Its Promise: A Multidisciplinary Introduction.* Boulder, CO: Westview Press, pp. 56–73.

Stichter, S. (1988) Women, employment and the family. In S. Stichter and J. Parpart (eds) *Women, Employment and the Family in the International Division of Labour.* London: Macmillan, pp. 11–71.

Stiglitz, J. (2002) *Globalization and its Discontents.* New York: W. W. Norton.

Stoll, D. (1990) *Is Latin America Turning Protestant? The Politics of Evangelical Growth.* Berkeley, CA: University of California Press.

Sweig, J. (2009) *Cuba: What Everyone Needs to Know.* Oxford: Oxford University Press.

Swenson, J. J., Carter, C. E., Domec, J. C., and Delgado, C. I. (2011) Gold mining in the Peruvian Amazon: Global prices, deforestation, and mercury imports. *PLoS ONE* 6(4): e18875.

Tana, K. (2010) Soybean wars. *COHA* 11 May. www.coha.org/soybean-wars (Accessed 6 November 2011).

Tatum, C. (1994) From Sandino to Mafalda: Recent works on Latin American popular culture. *Latin American Research Review* 29(1): 198–214.

Taylor, D. (2009) Evo Morales – From poverty to power. *Labour Campaign for International Development* 12 August. http://lcid.org.uk/2009/12/08/evo (Accessed 4 August 2011).

Thale, G. (2009) Beyond the Honduran coup. *Foreign Policy in Focus* 1 July. www.fpif.org/articles/behind_the_honduran_coup (Accessed 25 October 2011).

Tiano, S. (1990) Maquiladora women: A new category of workers. In K.Ward (ed.) *Women Workers and Global Restructuring.* New York: ILR Press, pp.193–223.

Tiano, S. and Ladino, C. (1999) Dating, mating and motherhood: Identity construction among Mexican maquila workers. *Environment and Planning A* 31(2): 305–325.

Tomlinson, J. (1991) *Cultural Imperialism: A Critical Introduction.* London: Pinter.

Travassos, C. and Williams D. R. (2004) The concept and measurement of race and their relationship to public health: A review focused on Brazil and the United States. *Cadernos de Saúde Pública* 20(3): 660–678.

Trigo, A. (2004) General introduction. In A. del Sarto, A. Ríos and A. Trigo (eds) *The Latin American Cultural Studies Reader.* Durham, NC: Duke University Press, pp. 1–14.

Trigona, M. (2004) The making of piquetero television. *NACLA Report on the Americas* 37(4): 32–33.

Trigona, M. (2007) Community television in Argentina: Ágora TV, a window for liberation. *Upside Down World* 11 June. http://upsidedownworld.org/main/argentina-archives-32/769-community-television-in-argentina-gora-tv-a-window-for-liberation (Accessed 20 January 2012).

Tuckman, J. (2007) Urban beach lets city's poor enjoy taste of Acapulco. *The Guardian* 5 April. www.guardian.co.uk/world/2007/apr/05/travelnews.mexico? (Accessed 12 November 2011).

Tuckman, J. (2011) Mexico's war on drugs blights resort of Acapulco. *The Guardian* 7 December. www.guardian.co.uk/world/2011/dec/07/mexico-drug-war-blights-acapulco (Accessed 27 March 2012).

Turner, T. (1991) The social dynamics of video media in an indigenous society: The cultural meaning and the personal politics of video-making in Kayapo communities. *Visual Anthropology Review* 7(2): 68–76.

Turner, T. (1992) Defiant images: The Kayapo appropriation of video. *Anthropology Today* 8(6): 5–16.

Turner, T. (2002) Representation, politics, and cultural imagination in indigenous video: General points and Kayapo examples. In F. D. Ginsburg, L. Abu-Lughod and B. Larkin (eds) *Media Worlds: Anthropology on New Terrain*. Berkeley, CA: University of California Press, pp. 75–89.

University of Texas (2007) *Unfulfilled Promises and Persistent Obstacles to the Realization of the Rights of Afro-Colombians: A Report on the Development of Ley 70 of 1993*. Report submitted to the Inter-American Commission on Human Rights. Austin, TX: University of Texas School of Law.

Urkidi, L. and Walter, M. (2011) Dimensions of environmental justice in anti-gold mining movements in Latin America. *Geoforum* 42(6): 683–695.

Valente, M. (2011) Argentina shows world how to beat the economic crisis. *Upside Down World* 19 December. http://upsidedownworld.org/main/argentina-archives-32/3366-argentina-shows-world-how-to-beat-the-economic-crisis (Accessed 27 March 2012).

Valenzuela, M. E. (1995) El programa nacional de apoyo a jefas de hogar de escasos recursos. In M. E. Valenzuela, S. Venegas and C. Andrade (eds) *De Mujer Sola a Jefa de Hogar: Género, Pobreza y Políticas Públicas*. Santiago, Chile: SERNAM, pp. 185–214.

van Cott, D. L. (2008) *Radical Democracy in the Andes*. Cambridge: Cambridge University Press.

van Cott, D. L. (2010) Indigenous peoples' politics in Latin America. *Annual Review of Political Science* 13: 385–405.

Vargas-Lundius, R., Lanly, G., Villareal, M. and Osorio, M. (2008) *International Migration, Remittances and Rural Development*. Rome, Italy: IFAD and FAO.

Varley, A. (1996) Women heading households: Some more equal than others? *World Development*, 24(3) 505–520.

Venegas, C. (2010) *Digital Dilemmas: The State, the Individual, and Digital Media in Cuba*. New Brunswick, NJ: Rutgers University Press.

Vidal, J. (2011) Bolivia enshrines natural world's rights with equal status for Mother Earth. *The Guardian* 10 April. www.guardian.co.uk/environment/2011/apr/10/bolivia-enshrines-natural-worlds-rights (Accessed 9 November 2011).

Vinter, H. (2010) Change gonna come: Gay rights in Latin America. *Argentina Independent* 19 April. www.argentinaindependent.com/currentaffairs/

newsfromlatinamerica/change-gonna-come-gay-rights-in-latin-america-
(Accessed 12 December 2011).

Vulliamy, E. (2012) Western banks "reaping billions from Colombian cocaine trade".
The Guardian 2 June. www.guardian.co.uk/world/2012/jun/02/western-banks-
colombian-cocaine-trade (Accessed 23 August 2012).

Wade, P. (1997) *Race and Ethnicity in Latin America.* London: Pluto Press.

Wade, P. (2002) Introduction: The Colombian Pacific in perspective. *Journal of Latin
American Anthropology* 7(2): 2–33.

Wade, P. (2006) Afro-Latin studies: Reflections on the field. *Latin American and
Caribbean Ethnic Studies* 1(1): 105–124.

Wadi, R. (2011) Living under the oppression of democracy – The Mapuche people of
Chile. *Upside Down World* 31 January. http://upsidedownworld.org/main/chile-
archives-34/2887-living-under-the-oppression-of-democracy-the-mapuche-people-
of-chile (Accessed 17 December 2011).

Wainwright, J. (2008) *Decolonizing Development: Colonial Power and the Maya.*
Malden, MA: Blackwell.

Wainwright, J. and Bryan, J. (2009) Cartography, territory, property: Postcolonial
reflections on indigenous counter-mapping in Nicaragua and Belize. *Cultural
Geographies* 16: 153–178.

Waisbord, S. (2000) *Watchdog Journalism in South America: News, Accountability
and Democracy.* New York: Columbia University Press.

Waisbord, S. (2002) Grandes gigantes: Media concentration in Latin America. *Open
Democracy* 27 February. www.opendemocracy.net/media-globalmedia
ownership/article_64.jsp (Accessed 12 January 2012).

Walsh, C. (2007) Shifting the geopolitics of critical knowledge: Decolonial thought
and cultural studies "others" in the Andes. *Cultural Studies* 21(2–3): 224–239.

Walsh, C. (2010) Development as Buen Vivir: Institutional arrangements and
(de)colonial entanglements. *Development* 53(1): 15–21.

Walsh, P. (2001) *Men Aren't from Mars: Unlearning Machismo in Nicaragua.*
London: CIIR.

Warf, B. (2008) *Time–Space Compression: Historical Geographies.* London:
Routledge.

Warren, J. (2001) *Racial Revolutions: Antiracism and Indian Resurgence in Brazil.*
Durham, NC: Duke University Press.

Wearden, G. (2011) Brazil considers helping Portugal ease debt crisis. *The Guardian*
30 March. www.guardian.co.uk/business/2011/mar/30/brazil-considers-helping-
portugal (Accessed 27 March 2012).

Wearne, P. (1994) *The Maya of Guatemala.* London: Minority Rights Group.

Wearne, P. (1996) *The Return of the Indian: Conquest and Revival in the Americas.*
Philadelphia, PA: Temple University Press.

Webster, D. L. (2002) *The Fall of the Ancient Maya: Solving the Mystery of the Maya
Collapse.* London: Thames and Hudson.

Weinberg, C. (2006) This is not a love story: Using soap opera to fight HIV in
Nicaragua. *Gender and Development* 14(1): 37–46.

Weisbrot, M. (2011) Commentary: Obama's Latin America policy: Continuity without
change. *Latin American Perspectives* 38(4): 63–72.

Wiarda, H. J. and Kline, H. F. (2011) The Latin American tradition and the process of
development. In H. J. Wiarda and H. F. Kline (eds) *Latin American Politics and
Development.* Boulder, CO: Westview Press, pp. 1–98.

Williamson, E. (2009) *The Penguin History of Latin America.* London: Penguin.

Wilpert, G. (2004) Community airwaves in Venezuela. *NACLA Report on the Americas* 37(4): 34–35.

Winacur, R. (2002) *Ciudadanos mediáticos: La construcción de lo público en la radio.* Barcelona: Gedisa.

Woods, N. (2006) *The Globalizers: The IMF, the World Bank, and Their Borrowers.* Ithaca, NY: Cornell University Press.

Yashar, D. J. (2005) *Contesting Citizenship in Latin America: The Rise of Indigenous Movements and the Postliberal Challenge.* Cambridge: Cambridge University Press.

Yúdice, G. (2003) *The Expediency of Culture: Uses of Culture in the Global Era.* Durham, NC: Duke University Press.

Zack-Williams, A. B. (2000) Social consequences of structural adjustment. In G. Mosha, E. Brown, B. Milward and A. B. Zack-Williams (eds) *Structural Adjustment: Theory and Practice.* London: Routledge.

Zamosc, L. (2004) The Indian movement in Ecuador: From politics of influence to politics of power. In N. G. Postrero and L. Zamosc (eds) *The Struggle for Indigenous Rights in Latin America.* Eastbourne, UK: Sussex Academic Press, pp. 131–157.

Zarsky, L. and Stanley, L. (2011) *Searching for Gold in the Highlands of Guatemala: Economic Benefits and Environmental Risks of the Marlin Mine.* Medford, MA: Tufts University, Global Development and Environment Institute. http://www.ase.tufts.edu/gdae/policy_research/marlinemine.pdf (Accessed 2 November).

Zibechi, R. (2011) Ecuador: The construction of a new model of domination. *Upside Down World* 5 August. http://upsidedownworld.org/main/ecuador-archives-49/3152-ecuador-the-construction-of-a-new-model-of-domination (Accessed 17 December 2011).

 # Index